Safe and Optimal Control of Safety-Critical Systems in Aerial and Space Robotics

Wasia Pia

Contents

Chapter 1

Introduction

Research in autonomous systems has boomed in the last decades. Fueled by the improvement in the available computational power on small devices, as well as by the need for intelligent and autonomous solutions in several different fields - ranging from assembly, medicine, and agriculture, as seen on Fig. 1.1 - we can hardly imagine our daily lives without them.

When we consider autonomous systems operating in high-risk scenarios, *safety* is of utmost importance. In this book, *safety* is defined as ensuring that an au-tonomous agent can navigate towards a goal while respecting state constraints

Figure 1.1: Autonomous systems cover a large part of our everyday lives: to the left, robotic manipulators assemble a car; to the top right, a robot is operated to perform a surgery; and on the bottom right, an agricultural robot patrols a field to maintain crops. Image rights: KUKA, DaVinci Surgery, GOFAR.

(avoiding prohibited areas, collisions, keeping a distance towards neighboring agents or objects) and actuation limits. The drive to automate many different parts of everyday life in the name of efficiency and cost reduction leads to the need to operate autonomous systems in many safety-critical scenarios, where previously only expert human operators were allowed to operate. Among such critical applications, we devote this work to two specific fields: the fields of aerial and space robotics.

Aerial technology - and in particular Unmanned Aerial Vehicles (UAVs) - has seen large developments towards applications in load transportation, inspection, surveillance, and swarm formations [1–6], Fig. 1.2. However, these are not all success stories; there have been reports of wide-scale crashes of UAVs transporting vital goods, causing disruptions in emergency services. This not only leads to technical and commercial losses, but also increases the disbelief in technology among the people that once it promised to help. One common problem during live operations that reportedly caused the crash of a Swiss Post drone[1] was disturbances that harshly affected the minimum viable performance of the system. In this case, the wind might have been too strong for the UAV to handle, leading to a large deviation from the planned trajectory and subsequent excitation of unsafe behaviors. As UAVs are increasingly operated near humans and may easily induce injuries, operators must completely avoid these situations. Simultaneously, energy consumption must be carefully monitored so that the operational time of these platforms - usually costly - can be extended. Therefore, control systems must generate optimal inputs considering the system properties and accounting for possible perturbations to ensure that the system is always operating within provably safe margins.

Moving from aerial to space robotics, we notice that Earth's orbit is getting increasingly crowded at each passing day. Motivated by the lowering costs of payload launch opportunities offered by commercial launch companies, such as SpaceX and Rocket Lab, an increasing number of institutions and companies are sending their own satellites that fulfil commercial and technology demonstration purposes. Applications for these systems may range from formation flying [7], human assistance [8], among others - see Fig. 1.3. However, and as big as Earth's orbit may be, safety issues arise somewhat regularly as the density of satellites in orbit increases. In 2019, a SpaceX Starlink and an European Space Agency satellites were set on a collision course that had to be manually solved by operators on ground control stations. As the number of small satellites (also known as CubeSats) is set to increase [9], such issues will happen more regularly and might have devastating consequences if not dealt with automatically, as shown by the number of debris that can result from a satellite collision [10].

To endow safety-critical systems with safe and optimal control algorithms that can automatically keep the system within safe operation margins, we address the following problem:

---

[1] IEEE Spectrum News Article (accessed on 18th of January, 2022): `https://spectrum.ieee.org/swiss-post-suspends-drone-delivery-service-after-second-crash`

Figure 1.2: Unmanned aerial vehicles find applications in collaborative and single agent aerial transportation (top left and right, respectively), maritime surveillance (bottom left) and inspection tasks (bottom right). Image rights: Swiss Post, Tekever, and Luleå University of Technology.

Problem 1: Consider an autonomous system operating in a shared environment. The goal is to design a control law that can safely track a trajectory that is provided by a high-level planning entity while using optimal control inputs for the system at hand and maintaining a safety margin from other agents or obstacles.

Independently of how well a single agent can perform on a given task, the redundancy and heterogeneity that a multi-agent system can achieve typically lead to more efficient and robust task completion. Good examples of such tasks are i) the load transportation scenario, where an oversized load might not be possible to be transported by a single agent but can be transported by a group of coordinated agents; and ii) the surveillance task, where multiple agents can cover more regularly the same area as compared to a single agent. In these scenarios, it is usually desirable that the agents keep a given relative position and orientation with respect to their neighbors to achieve an efficient load distribution or avoid overlapping surveilled areas. In the space systems context, satellite formations are employed for earth-positioning (GPS/GLONASS/Galileo constellations) and distributed aperture imaging systems. It is clear that there are numerous advantages to running a multi-agent setup, as long as safety is taken as assured since coordinating large groups of autonomous agents is still a topic of active research.

When considering cooperative behaviors, communication between the agents must exist either explicitly or implicitly. Each agent has to be aware of the relative localization of the nearby agents or know about its intent to safely and efficiently plan its motion. Typically, this involves exchanging state - and sometimes sensory

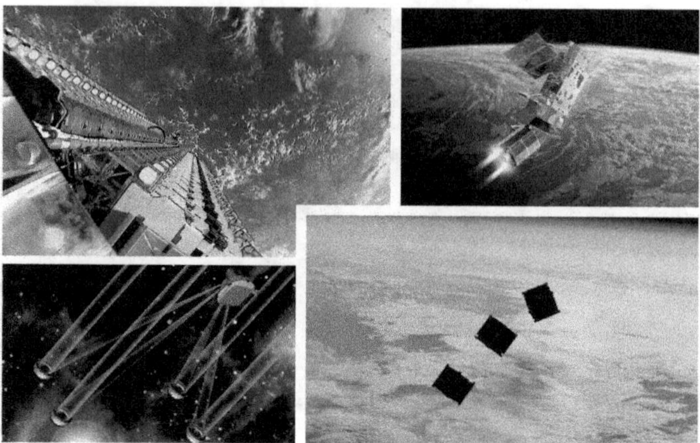

Figure 1.3: Space systems are booming, and satellite constellations are being used for communication (top left, Space-X Starlink constellation), deep space imaging (bottom left), de-orbiting space debris (top right) and perform research in orbital conditions (bottom right). Image rights: T. Herbst, Space-X, ESA and NASA.

- information, which might need a large bandwidth communication channel among all the involved parties. If such a communication channel ceases or its bandwidth is reduced, the group task may not be fulfilled. To deal with this problem and increase the robustness to the communication failure of autonomous formations, we set a goal to address Problem 2.

Problem 2: Consider multiple agents in a formation. The goal is to coordinate the multi-agent system by deriving local and distributed control laws, while minimizing the amount of shared information and considering the common formation objective.

A last but not minor component of this book deals with deriving control laws that take into account local sensing in each robot, particularly visual feedback. In aerial and space applications, autonomous agents are often deployed in environments where there is no global localization or reliable global orientation methods. The control task needs to be carried out using only locally sourced information in these scenarios.

Often, autonomous agents carry imaging systems for data collection and situational awareness, capable of detecting environment features of interest for several different purposes: inspection, scenario reconstruction, or state estimation. We could exploit the availability of this information to coordinate multiple robotic systems and dramatically extend the environments where we could deploy them and improve the performance of the already deployed ones. To explore this premise, we set a final goal to deal with the problem defined in Problem 3.

Problem 3: Consider an agent or a group of agents carrying on-board cameras. The goal is to control and coordinate each agent, or group of agents, considering visual information collected by generic imaging sensors.

We developed Robust and Safe Trajectory Tracking methods, Distributed Formation Control, and Image-based Coordination considering the problems above. We test these methods both on simulated and real systems. As a base-layer, we extensively use Model Predictive Control (MPC), as it allows us to derive optimal control inputs given known system dynamics and constraints. An overview of each chapter and its contributions is defined in the following section.

1.1 book **Outline and Contributions**

In this section, an outline of this book is provided, along with the contributions that each chapter relies on.

Chapter 2

This chapter deals with the preliminary knowledge needed for the scope of this book. Here we introduce visual servoing, epipolar geometry, model predictive control and control barrier functions, which will be used throughout the whole book. Suggested literature on these topics is also be provided.

Chapter 3

In this chapter, a framework for distributed model predictive control dedicated to formations of autonomous agents is presented - in this case, UAVs. This framework is motivated by the reduction of communicated data among the agents as a result of using a hierarchical leader-follower scheme, associated with novel predictive motion dynamics for followers that cooperate on maintaining a previously defined formation geometry. Simulation results shed light into the performance of such scheme considering static or time-varying references velocity references for the formation leader. This chapter is based on the following contribution:

1. Pedro Roque, Shahab Heshmati-Alamdari, Alexandros Nikou, Dimos V. Dimarogonas, "Decentralized Formation Control for Multiple Quadrotors under Unidirectional Communication Constraints", 21st IFAC World Congress (IFAC WC), Berlin, Germany, July 2020.

The methods presented in this chapter have been implemented and tested onboard the International Space Station on the NASA Astrobee robots in Fig. 1.4, on the 4th of December, 2021, and its results will be part of an upcoming publication.

Figure 1.4: Two Astrobees free-flying inside the International Space Station on an MIT/IST/KTH joint ReSWARM test session, on distributed coordination and control. Image rights: ESA and NASA.

In view of these test sessions, a number of contributions have been made to the open-source community, namely:

- Contributions to the Astrobee platform: Astrobee ROS Demo available at `https://github.com/Pedro-Roque/astrobee_ros_demo`, and direct contributions to the Astrobee flight software, available at `https://github.com/nasa/astrobee/commits?author=Pedro-Roque`

- Open-source Distributed MPC source code: available at `https://github.com/Pedro-Roque/reswarm_dmpc`

Chapter 4

In chapter 4, a formation control scheme is proposed for collaborative transportation, building on top of our previous work in [3] and [11]. In particular, we propose both a centralized and decentralized control scheme, based on Model Predictive Control, that steers the formation of UAVs towards a desired pose, while collaboratively transporting a load, considering linearization points of the whole system. Both simulations and experiments showcase the proposed solution. This chapter revolves on the following contribution:

2. Roberto Castro Sundin, Pedro Roque, Dimos V. Dimarogonas "Decentralized Model Predictive Control for Equilibrium-based Collaborative UAV Bar

Transportation", (accepted for publication at IEEE International Conference on Robotics and Automation (ICRA) 2022).

Chapter 5

Chapter 5 is dedicated to vision-based control. In the first part, a cascaded image-based visual servoing and model predictive controller are combined to optimally track a desired target speed for the UAV, given its dynamics. A stability proof for the MPC controller considering static setpoints is provided and experimental results shed insight into the performance of the suggested control law. Particularly, we are interested in exploring the separation of the two control methods, reducing the number of optimized variables to improve the solution time for the MPC controller. This chapter is based on the following contribution:

3. Pedro Roque, Elisa Bin, Pedro Miraldo and Dimos V. Dimarogonas, "Fast Model Predictive Image-Based Visual Servoing for Quadrotors", IEEE RSJ International Conference on Intelligent Robots and Systems (IROS), October 2020;

Chapter 6

In chapter 6, a MPC framework for image-based formation control is provided, capable of arranging multiple agents into desired relative poses based on image features and one distance measurement between any two agents in the constellation. Stability proofs for these controllers are provided, along with experimental results highlighting the performance of this approach. The following contribution constitutes the core of this chapter:

4. Pedro Roque, Pedro Miraldo, Dimos V. Dimarogonas, "Multi-Agent Relative Pose Coordination using Epipolar Constraints" (to be submitted to IEEE Transactions on Robotics (TRo)).

Chapter 7

In this chapter, a robust and safe control strategy is provided for time-varying trajectory tracking MPC. The suggested framework, denominated Corridor MPC, ensures that the system stays within a prescribe safe corridor around a provided trajectory by leveraging the safety properties of control barrier functions (CBFs) and the optimal control inputs provided by an MPC framework, given system dynamics and state and input constraints. This framework is applied to a freeflyer kinematic model in a simulation context. Stability proofs are provided for the closed-loop system. This chapter is based on the following contribution:

5. Pedro Roque, Wenceslao Shaw Cortez, Lars Lindemann, Dimos V. Dimarogonas, "Corridor MPC: Towards Optimal and Safe Trajectory Tracking", (accepted for publication at the American Control Conference (ACC) 2022);

Chapter 8

Lastly, chapter 8 concludes this book and sheds light into the future work that the current contributions might lead to.

Chapter 2

Notation and Preliminaries

In this chapter we will present the notation used throughout this book, as well as some fundamental concepts used on our contributions: model predictive control, zeroing control barrier functions, imaging systems, epipolar geometry and visual servoing.

First, let us address the notation used in this monograph. $\mathbb{N}$ is used for natural numbers, $\mathbb{N}_0$ for natural numbers including 0, $\mathbb{R}$ for real numbers, $\mathbb{R}_{>0}$ for all positive real numbers and $\mathbb{R}_{\geq 0}$ for the positive real numbers, including 0. We denote scalar and vector quantities as lower case $a \in \mathbb{R}^n$, and matrices as capital case $A \in \mathbb{R}^{n \times n}$, unless otherwise stated. An interval of values that includes a but excludes b is represented as $[a, b)$. All values belonging to the set $\mathbb{A}$ in the interval $[a, b]$ are represented as $\mathbb{A}_{[a,b]}$. The j-th component of a vector a is given by $a_{[j]}$. The set of a group of variables is defined as $\mathcal{A} = \{a, b, c\}$. The vector a represented in the coordinate system B is denoted as a^{B}. Similarly, rotation matrices mapping a vector in A to B are denoted as $R_{\mathsf{A}/\mathsf{B}}$, and relative position vectors as $t_{\mathsf{A}/\mathsf{B}}$.

We denote $\mathbb{A} \oplus \mathbb{B}$ and $\mathbb{A} \ominus \mathbb{B}$ as the Minkowski sum and Pontryagin difference of sets $\mathbb{A}$ and $\mathbb{B}$, respectively. The composition operator $\alpha \circ \beta$ represents $\alpha(\beta)$. The weighted norm of a vector $a \in \mathbb{R}^n$ by a positive definite matrix $A \in \mathbb{R}^{n \times n}$ is represented as $\|a\|_A = \sqrt{a^T A a}$. The identity and matrix in $\mathbb{R}^{n \times n}$ is defined as $I_{n \times n}$, while the zero matrix in $\mathbb{R}^{n \times m}$ is defined as $0_{n \times m}$, with $n, m \in \mathbb{N}$. The 2-norm of a vector $a \in \mathbb{R}^n$ is defined as $\|a\|_2$. The predicted value of b for time $(k + n)\Delta t$, predicted at time $k\Delta t$, are interchangeably written as $b(n\Delta t | k\Delta t)$ or $b(n|k)$, $k, n \in \mathbb{N}_0$. A continuous function, $\alpha : \mathbb{R} \to \mathbb{R}$ is an *extended class-$\mathcal{K}$ function* if it is strictly increasing and $\alpha(0) = 0$. A continuous function, $\beta : \mathbb{R} \to \mathbb{R}$ is a *class-$\mathcal{K}_\infty$ function* if it is strictly increasing, $\beta(0) = 0$ and $\lim_{r \to \infty} \beta(r) = \infty$. A continuous function, $\gamma : \mathbb{R} \times \mathbb{R} \to \mathbb{R}$ is a *class-$\mathcal{KL}$ function* if, for each fixed r, $\gamma(\cdot, r)$ is of class $\mathcal{K}$, and for each fixed s, $\gamma(s, \cdot)$ is decreasing and $\gamma(s, r) \to 0$ as $r \to \infty$.

2.1 Dynamics and Discretization

Let us first introduce the nonlinear control affine system

$$\dot{\tilde{x}} = f_{c,\tilde{x}}(\tilde{x}(t), u(t), w(t)) = f_c(\tilde{x}(t)) + g_c(\tilde{x}(t))u(t) + w(t), \tag{2.1}$$

where $\tilde{x}(t) \in \mathbb{R}^n$, $u(t) \in \mathbb{R}^m$, and $w(t) \in \mathbb{R}^n$ correspond to the state, the control input and piece-wise continuous disturbance, which evolve in their respective domains:

$$\tilde{x}(t) \in \tilde{\mathbb{X}}, \quad u(t) \in \mathbb{U}, \text{ and } w(t) \in \mathbb{W}. \tag{2.2}$$

The sets $\tilde{\mathbb{X}}$, $\mathbb{W}$ and $\mathbb{U}$ are assumed to be closed sets containing the origin. We further assume $f_c : \mathbb{R}^n \to \mathbb{R}^n$ and $g_c : \mathbb{R}^n \to \mathbb{R}^{n \times m}$ are locally Lipschitz continuous functions on $\tilde{\mathbb{X}}$. The disturbance w is unknown but bounded, with $w(t) \in \mathbb{W} := \{w \in \mathbb{R}^n : \|w\| \leq \overline{w}\}, \forall t \geq 0$. The discrete-time dynamics of (2.1) are represented as

$$\begin{aligned}
\tilde{x}((k+1)\Delta t) &= f_{\tilde{x}}(\tilde{x}, u, w) \\
&= f(\tilde{x}(k\Delta t)) + g(\tilde{x}(k\Delta t))u(k\Delta t) + w(k\Delta t), \tag{2.3}
\end{aligned}$$

for all $k \in \mathbb{N}_0$, and where f and g are the discrete-time dynamics of f_c and g_c in (2.1) obtained through an appropriate discretization method with sampling-time Δt. In the same way, the nominal dynamics for $x(k\Delta t) \in \mathbb{X}$ and considering $w(k\Delta t) = 0$, are given by

$$\begin{aligned}
x((k+1)\Delta t) &= f_x(x, u) \\
&= f(x(k\Delta t)) + g(x(k\Delta t))u(k\Delta t). \tag{2.4}
\end{aligned}$$

where $f_x(\tilde{x}, u) \triangleq f_{\tilde{x}}(\tilde{x}, u, 0)$.

2.2 Model Predictive Control

Dealing with safety-critical systems imposes several requirements on the control synthesis. In particular, three requirements must be met at all times: (i) ensure the system evolves along a safe set $\mathbb{X}_S \subset \tilde{\mathbb{X}}$; (ii) ensure that the actuation limits $\mathbb{U}$ are respected; and (iii) safely handle external disturbances $w \in \mathbb{W}$.

A common control approach to handle state and control constraints while minimizing the energy consumption is Nonlinear Model Predictive Control (NMPC). NMPC is a finite-horizon optimal controller (FHOC) that minimizes a cost function $J(x, u)$ along a receding horizon of length $N \in \mathbb{N}$ under state and control constraints, such as those in (2.2). The optimization problem is constrained by $x \in \mathbb{X}_S, u \in \mathbb{U}$ and the dynamics in (2.4). The optimization problem results in N predicted states and N control inputs for the system, of the form $\mathbf{x}_k^* = \{x^*(\Delta t|k\Delta t), \ldots, x^*(N\Delta t|k\Delta t)\}$, and $\mathbf{u}_k^* = \{u^*(0|k\Delta t), \ldots, u^*((N-1)\Delta t|k\Delta t)\}$ for a given initial state $x(0|k\Delta t) = \tilde{x}(k\Delta t)$, and an associated optimal cost value

$J^*(\tilde{x}(k\Delta t))$. Each discrete control input is applied to the system (2.1) in a Zero Order Hold (ZOH) fashion - a piece-wise constant input between sampling instances, that is, $u(t) = u^*(k\Delta t) \; \forall t \in [k\Delta t, (k+1)\Delta t)$. Noting that the predicted value of x for time $(k+n)\Delta t$, predicted at time $k\Delta t$, is $x(n\Delta t|k\Delta t), k, n \in \mathbb{N}_0$, the NMPC optimization problem is therefore defined as

$$J^*(\tilde{x}(k\Delta t)) = \min_{\mathbf{u}_k^*} \; J(x(n\Delta t|k\Delta t), u(n\Delta t|k\Delta t)) \tag{2.5a}$$

$$\text{s.t.:} \; x((m+1)\Delta t|k\Delta t) = f_x(x, u) \tag{2.5b}$$

$$x(m\Delta t|k\Delta t) \in \mathbb{X}_S \tag{2.5c}$$

$$u(m\Delta t|k\Delta t) \in \mathbb{U}, \; \forall m \in \mathbb{N}_{[0,N-1]} \tag{2.5d}$$

$$x(N\Delta t|k\Delta t) \in \mathbb{X}_F \subset \mathbb{X}_S \tag{2.5e}$$

$$x(0|k\Delta t) = \tilde{x}(k\Delta t).\forall n \in \mathbb{N}_{[0,N]} \tag{2.5f}$$

When solved at each sampling time $k\Delta t$, we obtain a feedback controller $K_N(\tilde{x}(k\Delta t)) = u^*(0|k\Delta t)$. The set $\mathbb{X}_F$ is a terminal control invariant set under a state feedback controller, such as $u_K(t) = Kx(t)$, for a given gain matrix K. It is common [12] to use Linear Quadratic Regulators (LQR) and associated control invariant sets as terminal sets in NMPC.

2.3 Zeroing Control Barrier Functions

Recently, Zeroing Control Barrier Functions (ZCBF) [13] have emerged as an alternative way to ensure safety of a continuous-time system (2.1). This formulation ensures forward invariance of a safe set where we wish the system to evolve, defining an RCI set. Let $h(\tilde{x}, t) : \mathcal{D} \times \mathbb{R}_{\geq 0} \to \mathbb{R}$, where $\mathcal{D} \subset \mathbb{R}^n$ is a compact set, be a continuously differentiable function, and let us associate with $h(\tilde{x}, t)$ the following sets

$$\mathcal{C}(t) = \{\tilde{x} \in \mathbb{R}^n : h(\tilde{x}, t) \geq 0\}, \tag{2.6a}$$

$$\partial\mathcal{C}(t) = \{\tilde{x} \in \mathbb{R}^n : h(\tilde{x}, t) = 0\}, \tag{2.6b}$$

where $\partial\mathcal{C}(t)$ is the boundary of $\mathcal{C}(t)$.

Definition 1 ([13]) *Given a set $\mathcal{C}(t)$ defined by (2.6) for a continuously differentiable function $h(\tilde{x}, t)$, then $h(\tilde{x}, t)$ is called a zeroing control barrier function defined on a set $\mathcal{D}$ with $\mathcal{C}(t) \subset \mathcal{D} \subset \mathbb{R}^n \; \forall t \geq 0$, if there exists a Lipschitz continuous extended class-$\mathcal{K}$ function α such that*

$$\sup_{u \in \mathbb{U}} \left[L_{f_c} h(\tilde{x}, t) + L_{g_c} h(\tilde{x}, t)u + \frac{\partial h(\tilde{x}, t)}{\partial t} + \alpha(h(\tilde{x}, t)) \right] \geq 0,$$

$$\forall \tilde{x} \in \mathcal{D}, \forall t \in \mathbb{R}_{\geq 0}, \tag{2.7}$$

where L_{f_c} and L_{g_c} are the Lie derivatives of h along f_c and g_c, respectively.

If $h(\tilde{x}, t)$ is a zeroing control barrier function and if we have a locally Lipschitz continuous control law $u(\tilde{x}, t)$ that satisfies the constraint $L_{f_c} h(\tilde{x}, t) + L_{g_c} h(\tilde{x}, t) u(\tilde{x}, t) + \frac{\partial h(\tilde{x}, t)}{\partial t} \geq -\alpha(h(\tilde{x}, t))$, which is as in (2.7), then the condition $\dot{h}(\tilde{x}(t), t) \geq -\alpha(h(\tilde{x}(t), t))$ is enforced for all $t \geq 0$ for the nominal system $\dot{\tilde{x}} = f_c(\tilde{x}(t)) + g_c(\tilde{x}(t)) u(t)$. By [13], it then follows that $\tilde{x}(t) \in \mathcal{C}(t)$ for all $t \geq 0$ when $\mathcal{C}(t)$ is compact.

It is worth remarking that the set $\mathcal{C}(t)$ can be seen as the set $\mathbb{X}_S$ in (2.5), with respect to the safety properties, and $\mathcal{D}$ as the set $\tilde{\mathbb{X}}$.

2.4 Imaging Systems

Current control applications often involve autonomous agents carrying onboard cameras to perform accurate localization and mapping tasks, as these are usually cheap sensors that can provide a large quantity of information. Assume that a robotic agent has a calibrated camera [14], where each world feature $p_i^{\mathsf{W}} := (X_i, Y_i, Z_i, 1)^T$ in the inertial frame, represented in homogeneous coordinates, is represented as a normalized image point $f_i^{\mathsf{C}} := (u_i, v_i, 1)^T$, where u_i and v_i are the normalized pixel coordinates in the camera frame, for each feature i. We model such cameras as a central projection system, which includes catadioptric cameras, perspective cameras, as well as several lens distortion models. Each projective ray $c_i^{\mathsf{C}} := \begin{bmatrix} x_i & y_i & z_i \end{bmatrix}^T$ - here abbreviated to c_i - is defined as $c_i = T_{\mathsf{W/C}} p_i^{\mathsf{W}}$, where the matrix $T_{\mathsf{W}}^{\mathsf{C}} \in \mathbb{R}^{3 \times 4}$ is the camera extrinsics matrix, parametrized by $T_{\mathsf{W/C}} = R_{\mathsf{C/W}}{}^T \begin{bmatrix} I & -t_{\mathsf{C/W}} \end{bmatrix}$, as in [14].

On what regards image formation, we consider two image formation models: the canonical perspective plane (CPP) [15] model, and the division (DIV) model [16]. We rely on the CPP model to represent image features in perspective (pin-hole), parabolic and hyperbolic cameras, whereas we use the division model for the distortion camera. These two models will be referred to as

$$f_i = \mathfrak{P}(c_i), \text{ and} \tag{CPP}$$
$$f_i = \mathfrak{D}(c_i) \tag{DIV}$$

Starting with the (CPP) model, the image ray c_i is first projected onto the unit sphere as $\bar{c}_i = c_i / \|c_i\| = \begin{bmatrix} \bar{x}_i & \bar{y}_i & \bar{z}_i \end{bmatrix}^T$. Then, the image feature f_i is obtained with

$$f_i = \frac{1}{\lambda} K_c \begin{bmatrix} \bar{x}_i \\ \bar{y}_i \\ \bar{z}_i + \xi \end{bmatrix} \tag{2.8}$$

where the parameter $\xi \in [0, 1]$ encodes the nonlinearities of general projective systems. Particular values for ξ are: $\xi = 0$ for perspective pin-hole cameras; $\xi = 1$ for parabolic cameras, and $\xi \in]0, 1[$ for hyperbolic cameras. Thus, this model allows us to represent multiple camera types under one imaging model. The parameter λ

is given by $\lambda = \bar{z}_i + \xi$. The matrix K_c represents the camera intrinsic parameters, defined as

$$K_c = \begin{bmatrix} a & 0 & c \\ 0 & b & d \\ 0 & 0 & 1 \end{bmatrix} \tag{2.9}$$

where a, b define the focal length of the camera, and c, d the optical center of the camera for X and Y axis, respectively.

Lastly, the (DIV) image model is used for cameras with lens distortion, such as fish-eye cameras. Features f_i are defined, in this model, according to

$$f_i = \frac{1}{z_i} K_c \begin{bmatrix} x_i(1 + \xi\sqrt{x_i^2 + y_i^2}) \\ y_i(1 + \xi\sqrt{x_i^2 + y_i^2}) \\ z_i \end{bmatrix} \tag{2.10}$$

for any $-1 \times 10^{-2} < \xi < 0$, as in [15]. In the forthcoming sections, we assume that the imaging model for each camera type is being employed and only refer to image features as f_i, independently of the camera topology, as it will be implicit.

Moreover, in this book we assume, without loss of generality, that all cameras are calibrated, and therefore we can replace $K_c = I_3$.

2.5 Epipolar Geometry

General projection systems epipolar constraint is defined by the nonlinear expression in [17, Eq. 27], in contrast to the linear constraint for perspective cameras, defined as

$$c_i^l E_j^l c_i^j = 0, \tag{2.11}$$

where j, l are two camera system frames and $j \neq l$, where E_j^l is the essential matrix. To avoid the nonlinear representation of the epipolar constraint for general camera systems, the image ray c_i is represented on a lifted coordinate space, as suggested in [15]. To this end, let the lifting of coordinates be obtained with the operator $\gamma : \mathbb{R}^3 \times \mathbb{R}^3 \to \mathbb{R}^6$, defined by

$$\gamma(c_i, c_{i+1}) = \begin{bmatrix} x_i x_{i+1} & \frac{x_i y_{i+1} + y_i x_{i+1}}{2} & y_i y_{i+1} & \frac{x_i z_{i+1} + z_i x_{i+1}}{2}, & \frac{y_i z_{i+1} + z_i y_{i+1}}{2}, & z_i z_{i+1} \end{bmatrix}^T \tag{2.12}$$

for two features i and $i+1$. With γ, a single point can also be lifted to the Veronese coordinate space, with

$$c_i \to {}_\diamond c_i = \gamma(c_i, c_i) = (x_i^2, x_i y_i, y_i^2, x_i z_i, y_i z_i, z_i^2)^T. \tag{2.13}$$

and back with

$$_\diamond c_i \to c_i = \gamma^{-1}({}_\diamond c_i) = (\sqrt{x_i^2}, \sqrt{y_i^2}, \sqrt{z_i^2})^T. \tag{2.14}$$

Similarly, the operator $\Lambda : \mathbb{R}^{3\times3} \to \mathbb{R}^{6\times6}$ is used to lift a linear operation Hx. Let $H = \begin{bmatrix} h_1 & h_2 & h_3 \end{bmatrix}$, where $h_1,\ldots,h_3$ are the columns of H, and $_\circ H$ the lifted representation of H. The operator Λ is defined as

$$_\circ H = \Lambda(H) = \begin{bmatrix} \gamma_{11} & \gamma_{12} & \gamma_{13} & \gamma_{22} & \gamma_{23} & \gamma_{33} \end{bmatrix} {}_\circ D \tag{2.15}$$

where $\gamma_{ij} = \gamma(h_i, h_j)$, and $_\circ D = diag\{1,2,1,2,2,1\}$.

To form the linear epipolar constraint for the various camera types in the lifted coordinate space, we use the results in [15, Tab. 2], presented in Table 2.1 for convenience. In this table, let

$$\Delta_c = \begin{bmatrix} 1-\xi^2 & 0 & 0 & 0 & 0 & -\xi^2 \\ 0 & 1-\xi^2 & 0 & 0 & 0 & 0 \\ 0 & 0 & 1-\xi^2 & 0 & 0 & -\xi^2 \\ 0 & 0 & 0 & 1 & 0 & 0 \\ 0 & 0 & 0 & 0 & 1 & 0 \\ 0 & 0 & 0 & 0 & 0 & 1 \end{bmatrix}, \tag{2.16}$$

$$\Theta = \begin{bmatrix} 0 & 0 & 0 & 2 & 0 & 0 \\ 0 & 0 & 0 & 0 & 2 & 0 \\ -1 & 0 & -1 & 0 & 0 & 1 \end{bmatrix}^T, \tag{2.17}$$

$$\Phi = \begin{bmatrix} 0 & 0 & 0 & 1 & 0 & 0 \\ 0 & 0 & 0 & 0 & 1 & 0 \\ \xi & 0 & \xi & 0 & 0 & 1 \end{bmatrix}^T, \tag{2.18}$$

$$\tag{2.19}$$

With the generalized essential matrices in Table 2.1, we can write the general epipolar constraint in the lifted space as

$$_\circ c^l \, \tilde{E}^l_j \, {}_\circ c^j = 0. \tag{2.20}$$

2.6 Visual Servoing

Image Based Visual Servoing (IBVS) [18] consists of controlling a camera movement based solely on image feature observations. Consider an observed feature f_i and corresponding desired feature position in the camera frame, $\bar{f}_i$. The goal of the visual servoing task is to drive the robot to a position in which f_i converges to $\bar{f}_i$, that corresponds to the desired 3D position of the robotic agent. The error between the desired and current feature observations is defined as

$$\tilde{f}_i = f_i - \bar{f}_i \;\Rightarrow\; \dot{\tilde{f}}_i = \dot{f}_i. \tag{2.21}$$

Next, we introduce the interaction matrix $L(f_i, \xi, \rho) \in \mathbb{R}^{2\times6}$, referred to as L_i, that relates the velocity of a generalized camera in a robotic agent, $u \in \mathbb{R}^6$, with

Camera 1 - Camera 2	Generalized Essential Matrix - $\tilde{E}_j^l$
Pin-hole - Pin-hole	$\underbrace{\begin{bmatrix} 0_{3\times 3} & 0_{3\times 3} \\ 0_{3\times 3} & E_j^l \end{bmatrix}}_{\vee E_j^l}$
Pin-hole - Hyperbolic	$_\diamond D \,_\diamond E_j^l \Delta_c{}^T$
Pin-hole - Parabolic	$_\vee E_j^l \Theta^T$
Pin-hole - Distortion	$_\vee E_j^l \Phi^T$
Parabolic - Parabolic	$\Theta \,_\vee E_j^l \Theta^T$
Parabolic - Distortion	$\Theta \,_\vee E_j^l \Phi^T$
Distortion - Distortion	$\Phi \,_\vee E_j^l \Phi^T$

Table 2.1: Generalized essential matrices for the epipolar constraint in the lifted Veronese coordinate space.

the movement of the observed features:

$$\begin{bmatrix} \dot{u}_i \\ \dot{v}_i \end{bmatrix} = L_i u, \tag{2.22}$$

where $L_i \in \mathbb{R}^{2\times 6}$ [19, Eq. 13] is defined as

$$L_i = \begin{bmatrix} -\dfrac{1+u_i^2(1-\xi(\alpha+\xi))+v_i^2}{\rho(\alpha+\xi)} & \dfrac{\xi u_i v_i}{\rho} & \dfrac{\alpha u_i}{\rho} \\[2ex] \dfrac{\xi u_i v_i}{\rho} & -\dfrac{1+u_i^2(1-\xi(\alpha+\xi))+v_i^2}{\rho(\alpha+\xi)} & \dfrac{\alpha v_i}{\rho} \\[3ex] u_i v_i & -\dfrac{(1+u_i^2)\alpha-\xi v_i^2}{\alpha+\xi} & v_i \\[3ex] -\dfrac{(1+v_i^2)\alpha-\xi u_i^2}{\alpha+\xi} & -u_i v_i & -u_i \end{bmatrix}, \tag{2.23}$$

with $\alpha = \sqrt{1 + (1 - \xi^2)(u_i^2 + v_i^2)}$ and $\rho = \sqrt{x_i^2 + y_i^2 + z_i^2}$. For the case of a perspective camera ($\xi = 0$), (2.23) becomes

$$L_i = \begin{bmatrix} -\frac{1}{z_i} & 0 & \frac{u_i}{z_i} & u_i \cdot v_i & -(1 + u_i^2) & v_i \\ 0 & -\frac{1}{z_i} & \frac{v_i}{z_i} & 1 + v_i^2 & -u_i \cdot v_i & -u_i \end{bmatrix}. \tag{2.24}$$

Note that, in this case, u is composed by a linear and an angular velocity components, $\nu \in \mathbb{R}^3$ and $\omega \in \mathbb{R}^3$, respectively, such that

$$u = \begin{bmatrix} \nu^T & \omega^T \end{bmatrix}^T. \tag{2.25}$$

From (2.24), features with $z_i = 0$ cause L_i to be ill-defined. This happens when points are in a plane that coincides with the image plane, passing through the origin. In this case, the points will not be visible in the image. Therefore, L_i is well-defined for the considered operating range. It is important to note that (2.23) requires an estimate of the depth z_i. This can be achieved with onboard monocular estimators, as shown in the literature [20].

Chapter 3

Formation Control under Unidirectional Communication

After participating in three test sessions on the International Space Station with the MIT SPHERES robots [21], we observed that communication between multiple agents could, sparsely, fail. In this test session, we relied on predicted control inputs from the NMPC controller to account for communication loss - in case of communication failure on a given sampling instance, we would use the previously predicted control input for a given predicted time step, relying on the receding horizon nature of the controller until no more predicted inputs are available. This observation led to the following question: *could we develop a NMPC approach that requires less communication, since all agents know a-priori that they belong to a specific geometric configuration?* The answer to this question lies in this chapter, as a proposed solution to formation control under unidirectional communication constraints in a Leader-Follower scheme.

3.1 Literature Review

During the last decades, considerable progress has been made in the field of UAVs, with a significant number of results in a variety of aerial surveillance activities, as in [22].

Classic approaches such as feedback linearization in [23], Lyapunov based stabilization in [24], dynamic inversion in [25], PID and LQR in [26] and backstepping in [27] have been used in the past to design motion controllers for rotary wing aerial robots. Nevertheless, the aforementioned methods yield poor closed-loop performance and the results were local, around only selected operating points. In addition, the aforementioned motion control strategies do not consider input and state constraints. In this context, NMPC [28] is a suitable approach for complex surveillance missions, as it is able to combine motion planning with input and state constraints satisfaction, as in [29].

Figure 3.1: Representation of the UAV model.

Motivated by the aforementioned considerations, this chapter presents a novel cooperative control framework for a group of multiple UAVs in order to follow a desired path while maintaining a predefined formation geometry and satisfying input and state constraints. In particular, we propose a novel decentralized leader-follower architecture, where the leader UAV, which has knowledge over a predefined formation geometry and desired path, tries to navigate the whole group of agents while ensuring the connectivity of the team. More specifically, in order to maintain the connectivity, the leader UAV estimates the followers' motion, by employing a fast geometric propagation that exploits the knowledge of the desired formation and relative position sensing information. On the other hand, the followers estimate the motion of the leader UAV, by receiving its local state information and performing a first order Euler propagation. A NMPC law that tracks the desired formation and maintains the connectivity with respect to the leader UAV is implemented in each follower. Specifically, no explicit data is sent from the followers to other agents involved in the cooperative task, making the proposed approach scalable regarding the number of agents involved in the formation task.

3.2 Problem Formulation

Consider $m + 1$ aerial robotic agents under a single leader and multiple followers architecture operating. We denote the agents state vectors by x_i, $i \in \mathcal{A} = \{\mathsf{L}, \mathsf{F}_1, \ldots, \mathsf{F}_M\}$, where L represents the leader, $\mathsf{F}_1 \ldots \mathsf{F}_M$ the followers. Let $\mathcal{F} = \mathcal{A}\backslash\mathsf{L}$. The state of each agent, defined as $x_i \triangleq [p_i, v_i, q_i, \omega_i]^T \in \mathbb{R}^3 \times \mathbb{R}^3 \times \mathbb{S}^3 \times \mathbb{R}^3$, $i \in \mathcal{A}$, includes the position (i.e., $p_i \in \mathbb{R}^3$), linear velocity (i.e., $v_i \in \mathbb{R}^3$), attitude quaternion (i.e., $q_i \in \mathbb{S}^3$), and angular velocity ($\omega_i \in \mathbb{R}^3$) vectors for agent i. Unless otherwise stated, x_i and v_i are represented in the inertial frame I, while q_i and ω_i are represented in the body frame of agent $i \in \mathcal{A}$. Relative positions in the agents' body frame are defined as χ_{ij}, and represent the position of agent j, in the frame of agent i. It is often more practical to use rotation matrices that are derived from the attitude quaternion q_i. Such matrices are represented as $R_{i/b}(q_i) : \mathbb{S}^3 \to \mathbb{SO}(3)$ and rotate a vector from frame i to frame b. We omit the target from when the we consider the inertial frame I, as in $R_{i/I} = R_i$. All rotation matrices have the property $||R_i|| = 1$ and $\det(R_i) = 1$. Without loss of generality, according to the standard aerial vehicle's modeling properties in [30], the dynamic equations of each agent $i \in \mathcal{A}$ can be given as:

$$\dot{p}_i = v_i, \tag{3.1a}$$

$$\dot{v}_i = R_i \frac{\nu_i}{m_i} + g, \tag{3.1b}$$

$$\dot{q}_i = \frac{1}{2}\Omega(q_i)\omega_i, \tag{3.1c}$$

$$\dot{\omega}_i = M_i^{-1}(\tau_i - \omega_i \times M_i\omega_i), \tag{3.1d}$$

where m_i is the vehicle's mass, g the gravity vector expressed in the inertial frame I, M_i the vehicle's inertia matrix, $\tau_i \in \mathbb{R}^3$ is the 3-DoF (Degree-of-Freedom) torque vector in the body frame, and $\Omega(q_i)$ is defined as:

$$\Omega(q_i) = \begin{bmatrix} q_{i[w]} & -q_{i[z]} & q_{i[y]} \\ q_{i[z]} & q_{i[w]} & -q_{i[x]} \\ -q_{i[y]} & q_{i[x]} & q_{i[w]} \\ -q_{i[x]} & -q_{i[y]} & -q_{i[z]} \end{bmatrix}.$$

Moreover, $\nu_i \triangleq [0 \ \ 0 \ \ \nu_{i[z]}]^T$ represents a 1-DoF thrust aligned with the z axis of the agent body frame. From now on, we denote by $u_i \triangleq \left[\nu_{i[z]}, \tau_i^T\right]^T \in \mathbb{R}^4$ the concatenated control input of the robotic agent i. Based on the aforementioned considerations, the dynamics of (3.1a)-(3.1d), can be written in discrete-time form as:

$$x_i(n+1|k) = f_i(x_i(n|k), u_i(n|k)), \ i \in \mathcal{A}, \tag{3.2}$$

where $f_i : \mathbb{R}^6 \times \mathbb{S}^3 \times \mathbb{R}^3 \times \mathbb{R}^4 \to \mathbb{R}^6 \times \mathbb{S}^3 \times \mathbb{R}^3$ defines the discrete-time dynamics for (3.1a)-(3.1d), discretized through a zero-order-hold (ZOH) sampling method. Assume that the velocity that the UAV can generate is bounded by $v_{i_{Max}}$, $i \in \mathcal{A}$. Moreover, quaternions are unit-norm vectors. These requirements are captured by the state constraint set Ξ_i, given by

$$\Xi_i \triangleq \{x_i \in \mathbb{R}^9 \times \mathbb{S}^3 : ||v_i|| \leq v_{i_{Max}}, ||q_i|| = 1\}, \ i \in \mathcal{A}. \tag{3.3}$$

The actuation forces and torques are generated by the thrusters. Thus, we define the control constraint set U_i, $i \in \mathcal{A}$, corresponding to the physical actuator limits, as follows:

$$U_i \triangleq \Big\{ u_i \in \mathbb{R}^4 : -\frac{\nu_{i[z]}}{4}a_{l_i} \leq \tau_{[x,y]} \leq \frac{\nu_{i[z]}}{4}a_{l_i},$$
$$-0.01 \leq \tau_{[z]} \leq 0.01, 0 \leq \nu_{i[z]} \leq 4 \cdot 9.81 \cdot m_i \Big\}, \tag{3.4}$$

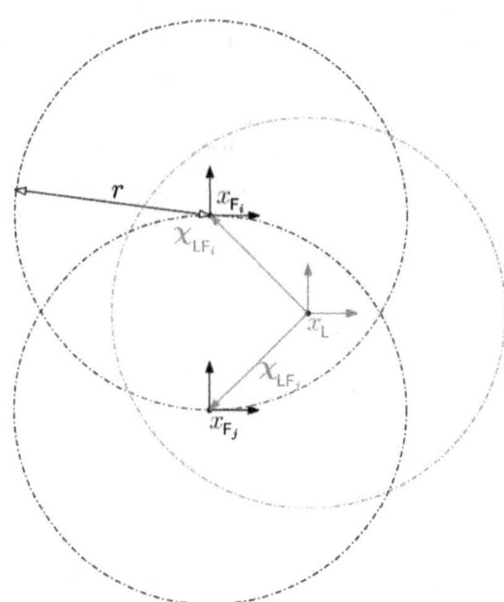

Figure 3.2: Connectivity for a team of agents including one leader and two followers. The sensing areas are denoted by $\mathcal{B}(x_i, r)$, $i \in \mathcal{A}$.

where a_{l_i} is the quadrotor's arm length, i.e., the distance between the center of the UAV and the motors, as in Fig. 3.1.

Consider the set of agents $\mathcal{A}$. Without loss of generality, let $\mathcal{B}(x_i, r)$ be a sensing area where the agent $i \in \mathcal{A}$ can measure a relative distance d_{ij} and a unitary relative bearing vector β_{ij} with respect to its neighbor $j \in \mathcal{N}_i$, where

$$\mathcal{N}_i \triangleq \{j \in \mathcal{A}, j \neq i : ||p_i - p_j|| \leq r\}. \tag{3.5}$$

It should be noted that $\mathcal{N}_i$ denotes the agents that are required to remain connected, i.e. within the sensing radius of its neighbors, (see Fig 3.2). The aforementioned values can be defined as

$$d_{ij} = ||p_j - p_i||, \quad \beta_{ij} = R_i^T \frac{p_j - p_i}{d_{ij}}, \quad \chi_{ij} = d_{ij} \cdot \beta_{ij}, \tag{3.6}$$

where $d_{ij} > 0$. Note that from a range d_{ij} and bearing β_{ij} we can uniquely reconstruct χ_{ij} - the relative position of agent j in the frame of i - and that these measurements can be obtained locally, therefore requiring no communication. Accordingly, the formation geometry is defined as

$$\hat{\chi}_{ij} = \hat{d}_{ij}\,\hat{\beta}_{ij}, \quad i \in \mathcal{F}, j \in \mathcal{N}_i, \tag{3.7}$$

where $\hat{d}_{ij}$, $\hat{\beta}_{ij}$ and $\hat{\chi}_{ij}$ are desired relative distances, bearings and positions. Note that each follower $i \in \mathcal{F}$ is only aware of its own desired relative distance, bearing and position with respect to the leader UAV.

Moreover, we consider that each follower i, $i \in \mathcal{F}$ receives a set of information broadcasted by the leader, denoted as

$$w_{\mathsf{L}} = [x_{\mathsf{L}}^T, \hat{v}_{\mathsf{L}}^T]^T \in \mathbb{R}^6 \times \mathbb{S}^3 \times \mathbb{R}^3 \times \mathbb{R}^3 \subset \mathcal{W} \tag{3.8}$$

where $\hat{v}_{\mathsf{L}}$ is the reference leader velocity on the desired path [23, 31] and $\mathcal{W}$ is a compact set capturing the information broadcasted by the leader and satisfies $\mathcal{W} \subset \Xi_{\mathsf{L}}$.

The control objective is to guide the team of agents to follow cooperatively a desired path which is known only to the formation leader, while guarantying i) the maintenance of a predefined formation geometry, ii) the connectivity among all agents, as well as iii) respecting state and input constraints. Hence, the problem addressed in this chapter is formally written as:

Problem 1 *Consider the set $\mathcal{A}$ of UAVs described by (3.2), with state and input constraints imposed by the sets Ξ_i, U_i, $i \in \mathcal{A}$, given in (3.3) and (3.4), respectively. Considering that i) only the leader UAV is aware of a desired path, and ii) the followers have access only to information broadcasted by the leader UAV defined in (3.8), design a decentralized control protocol u_i, $i \in \mathcal{A}$, that imposes the team of agents to follow cooperatively the desired path p_d while guaranteeing i) the maintenance of the predefined formation geometry in (3.7), and ii) the connectivity among the agents with respect to their sensing capabilities defined in (3.5).*

3.3 Control Methodology

We propose a control strategy that requires the desired path p_d to be known and tracked by the leader. The leader UAV navigates the whole team by keeping the whole group connected and organized according to the desired relative positions (3.7). We propose a decentralized NMPC approach to solve Problem 1.

Control Strategy

At each sample time k, a cost function $J_i(\cdot)$, $i \in \mathcal{A}$, is minimized with respect to a control sequence $\mathbf{u}_i^* = \{u_i(0|k), u_i(1|k), \ldots, u_i(N-1|k)\}$, yielding N predicted states $\{x_i(1|k), \ldots, x_i(N|k)\}$ by applying the optimal control input to the agents dynamics. We also have that $x_i(0|k) = x_i(k)$, corresponding to the actual feedback of agent i at time k. Depending on the agent's role - either leader or follower - different assumptions and costs $J_i(\cdot)$ are defined.

Error Definition

The vector of errors for the leader UAV as well as follower UAVs are, respectively,

$$
e_{\mathsf{L}} = \begin{bmatrix} e_{\mathsf{LF}} \\ e_{v_{\mathsf{L}}} \\ e_{R_{\mathsf{L}}} \end{bmatrix} = \begin{bmatrix} \sum_{i \in \mathcal{N}_{\mathsf{L}}} (\hat{\chi}_{\mathsf{L}i} - \chi_{\mathsf{L}i}) \\ \hat{v}_{\mathsf{L}} - v_{\mathsf{L}} \\ e_{R_{\mathsf{L}}}(R_{\mathsf{L}}, \hat{R}_{\mathsf{L}}) \end{bmatrix} \in \mathbb{R}^7, \tag{3.9a}
$$

$$
e_i = \begin{bmatrix} e_{i\mathsf{L}} \\ e_{v_i} \\ e_{R_i} \end{bmatrix} = \begin{bmatrix} \hat{\chi}_{i\mathsf{L}} - \chi_{i\mathsf{L}} \\ \hat{v}_i - v_i \\ e_{R_i}(R_i, \hat{R}_i) \end{bmatrix} \in \mathbb{R}^7, \; i \in \mathcal{F}, \tag{3.9b}
$$

where $\hat{v}_i$ is the reference leader velocity (i.e., $\hat{v}_{\mathsf{L}}$ broadcasted by the leader) transformed to the followers body frame, $\hat{R}_i$, $i \in \mathcal{F}$, is a predefined desired attitude for the agents in the formation, and $e_{R_i}(R_i, \hat{R}_i)$ the same as of [32, Eq. 8], $e_{R_i}(R_i, \hat{R}_i) = \frac{1}{2}tr(I - \hat{R}_i^T R_i)$, where $tr(\cdot)$ is the trace operator. Note that $e_{R_i}(R_i, \hat{R}_i) \in [0, 2)$. The feasible error sets are defined as:

$$
E_{\mathsf{L}} = \{ e_{\mathsf{L}} \in \mathbb{R}^7 : \{\{\hat{\chi}_{\mathsf{L}i}\} \ominus \{\Xi_i \ominus \Xi_{\mathsf{L}}\}\} \cup \{\{\hat{u}\} \ominus \{u\}\} \cup [0,2) \}, \tag{3.10a}
$$

$$
E_i = \{ e_i \in \mathbb{R}^7 : \{\{\hat{\chi}_{i\mathsf{L}}\} \ominus \{\Xi_{\mathsf{L}} \ominus \Xi_i\}\} \cup \{\{v_i\} \ominus \{\hat{u}_i\}\} \cup [0,2) \}. \tag{3.10b}
$$

Leader Control Framework

The following section defines the controller for the formation leader, as well as the propagation method used to predict the followers motion.

We assume that the leader has access to its state x_{L}, relative distance $d_{\mathsf{L}i}$ and bearing $\beta_{\mathsf{L}i}$ measurements, with corresponding desired values $\hat{d}_{\mathsf{L}i}$ and $\hat{\beta}_{\mathsf{L}i}$, for all $i \in \mathcal{N}_i$, given in (3.7). Note that $d_{\mathsf{L}i} = d_{i\mathsf{L}}$, but $\beta_{\mathsf{L}i} \neq \beta_{i\mathsf{L}}$, due to (3.6), which results in $\chi_{\mathsf{L}i} \neq \chi_{i\mathsf{L}}$. We further consider a prediction function $\Upsilon_{e_i}(\cdot)$, which will be defined hereafter, predicting the relative position of the followers for a receding horizon of length N, based only on the measurements $\chi_{\mathsf{L}i}(k) = d_{\mathsf{L}i}(k)\beta_{\mathsf{L}i}(k)$ obtained at time k.

Now we are ready to formulate the NMPC problem for the leading agent L as

$$
\min_{\mathbf{u}_{\mathsf{L}}^*} J_{\mathsf{L}}(x_{\mathsf{L}}(k)) = \min_{\mathbf{u}_{\mathsf{L}}^*} \sum_{n=0}^{N-1} \left[V_P\big(e_{v_{\mathsf{L}}}(n|k), u_{\mathsf{L}}(n|k)\big) + \right.
$$
$$
V_Q\big(e_{\mathsf{LF}}(n|k)\big) + V_A\big(e_{R_{\mathsf{L}}}(n|k)\big) \Big] +
$$
$$
V_V\big(e_{\mathsf{L}}(N|k)\big). \tag{3.11}
$$

subject to

$$
x_{\mathsf{L}}(n+1|k) = f_{\mathsf{L}}\big(x_{\mathsf{L}}(n|k), u_{\mathsf{L}}(n|k)\big)
$$
$$
e_{\mathsf{LF}}(n+1|k) = \sum_{i \in \mathcal{N}_{\mathsf{L}}} \Upsilon_{e_i}\big(\hat{\chi}_{\mathsf{L}i} - \chi_{\mathsf{L}i}(n|k), v_{i_{\mathsf{Max}}}\big)
$$
$$
e_{\mathsf{L}}(n|k) \in E_{\mathsf{L}}, \quad e_{\mathsf{L}}(N|k) \in E_{\mathsf{L}}^f
$$
$$
u_{\mathsf{L}}(n|k) \in U_{\mathsf{L}} \quad , n = 0, \ldots, N-1,
$$

where

$$V_P\big(e_{v_L}(n|k), u_L(n|k)\big) =$$
$$= \|e_{v_L}(n|k)\|^2_{Q_V} + \|u_L(n|k)\|^2_{Q_R}, \tag{3.12a}$$
$$V_Q\big(e_{LF}(n|k)\big) = \|e_{LF}(n|k)\|^2_{Q_F}, \tag{3.12b}$$
$$V_A\big(e_{R_L}(n|k)\big) = \|e_{R_L}(n|k)\|^2_{Q_A}, \tag{3.12c}$$
$$V_V\big(e_L(n|k)\big) = \|e_L(n|k)\|^2_{Q_N}, \tag{3.12d}$$

and where $E_L^f \subset E_L$ is the terminal set, $v_{i_{Max}}$ the maximum velocity for all $i \in \mathcal{N}_i$ given in (3.3), and Q_F, Q_V, Q_R, Q_A and Q_N are positive definite diagonal weighing matrices (Q_A is a scalar). The cost functions (3.12a) to (3.12c) represent the running cost for the velocity error, relative position error, attitude error and control input, while (3.12d) represents the terminal cost.

Follower Position Propagation

Let $\chi_{Li}(k)$ be the relative position of agent i in the frame of the formation leader, at time k. This position is calculated based on the leader measurements i.e, $d_{Li}(k)$ and $\beta_{Li}(k)$ through equation (3.6).

Taking into account the state constraints in (3.3), a ball $\mathcal{B}(\chi_{Li}(n|k), \Delta v_{i_{Max}}\Delta t)$ defines the set of all possible relative positions $\chi_{Li}(n+1|k)$ where $\Delta v_{i_{Max}} = v_{i_{Max}} - \|v_L(n|k)\|$, and Δt is a constant sampling interval; $\mathcal{B}(\chi_{Li}(k), \Delta v_{i_{Max}}\Delta t)$ represents the set of all possible positions for the neighbor i over one step of the controller.

The formation leader, acting as a central coordinator in the formation setting, assumes that the followers strive to achieve their desired positions as fast as possible, navigating at maximum velocity to do so. However, as the movement is relative to the leader, its velocity also influences the set of possible positions to be achieved by the followers. The relative position propagation of the followers in the leader frame through the whole receding horizon is then predicted according to

$$\Upsilon_i(\chi_{Li}(n|k), \hat{\chi}_{Li}, v_{i_{Max}}) = \chi_{Li}(n|k) +$$
$$+ \frac{\hat{\chi}_{Li} - \chi_{Li}(n|k)}{\|\hat{\chi}_{Li} - \chi_{Li}(n|k)\|}(v_{i_{Max}} - \|v_L\|) \cdot (1 - e^{-\alpha \cdot \|\hat{\chi}_{Li} - \chi_{Li}(n|k)\|}) \cdot \Delta t, \tag{3.13}$$

where α is a tuning parameter that adjusts the speed gradient of the followers prediction, and e^x the exponential of x.

Finally, it is convenient to define the function Υ_i in terms of the error e_{LF}, so that we can include it on the NMPC formulation:

$$e_{\mathsf{LF}}(n|k) = \sum_{i \in \mathcal{N}_{\mathsf{L}}} \left(\hat{\chi}_{\mathsf{L}i} - \chi_{\mathsf{L}i}(n|k) \right) \tag{3.14}$$

$$e_{\mathsf{LF}}(n+1|k) =$$

$$= \sum_{i \in \mathcal{N}_{\mathsf{L}}} \hat{\chi}_{\mathsf{L}i} - \Upsilon_i(\chi_{\mathsf{L}i}(n|k), \hat{\chi}_{\mathsf{L}i}, v_{i_{\mathsf{Max}}}) =$$

$$= \sum_{i \in \mathcal{N}_{\mathsf{L}}} \Upsilon_{e_i}(\chi_{\mathsf{L}i}(n|k), \hat{\chi}_{\mathsf{L}i}, v_{i_{\mathsf{Max}}}). \tag{3.15}$$

Followers Control Framework

The follower's objective is to maintain a desired relative position with respect to the leader. The followers $i \in \mathcal{N}_{\mathsf{L}}$ do not possess any knowledge over the desired formation path, which is known only to the formation leader. The leader, then, broadcasts through $w_{\mathsf{L}}(k)$ the velocity that it tracks at time k, which the followers use to unify the formation movement. This means that the desired tracking velocity is kept constant for all followers along the receding horizon, whereas the leader can update this desired velocity based on its predicted states. Accordingly, we assume that the followers have access to their state x_i, the relative distance $d_{i\mathsf{L}}$ and bearing $\beta_{i\mathsf{L}}$ measurements, as well as to the desired formation constraints, $\hat{d}_{i\mathsf{L}}$ and $\hat{\beta}_{i\mathsf{L}}$. We are ready to define the NMPC for the followers as:

$$\min_{\mathbf{u}_i^*} J_i(x_i(k)) = \min_{\mathbf{u}_i^*} \sum_{n=0}^{N-1} \left[V_P\left(e_{v_i}(n|k), u_i(n|k)\right) + \right.$$

$$+ V_Q\left(e_{i\mathsf{L}}(n|k), w_{\mathsf{L}}(k)\right) + V_A\left(e_{R_i}(n|k)\right) \bigg] +$$

$$+ V_V\left(e_i(N|k)\right). \tag{3.16}$$

subject to

$$x_i(n+1|k) = f_i\big(x_i(n|k), u_i(n|k)\big)$$

$$e_{i\mathsf{L}}(n+1|k) = g_{e_{\mathsf{L}}}\big(e_{i\mathsf{L}}(n|k), w_{\mathsf{L}}(k)\big)$$

$$e_i(n|k) \in E_i, u_i(n|k) \in U_i \ , n = 0, \dots, N-1$$

$$e_i(N|k) \in E_i^f, \quad w_{\mathsf{L}}(k) \in W$$

where

$$V_P\left(e_{v_i}(n|k), u_i(n|k)\right) =$$

$$= \|e_{v_i}(n|k)\|_{Q_V}^2 + \|u_i(n|k)\|_{Q_R}^2, \tag{3.17a}$$

$$V_Q\left(e_{i\mathsf{L}}(n|k)\right) = \|e_{i\mathsf{L}}(n|k)\|_{Q_F}^2, \tag{3.17b}$$

$$V_A\left(e_{R_i}(n|k)\right) = \|e_{R_i}(n|k)\|_{Q_A}^2, \tag{3.17c}$$

$$V_V\left(e_i(n|k)\right) = \|e_i(n|k)\|_{Q_N}^2, \tag{3.17d}$$

and where $E_i^f \subset E_i$ is the terminal set and Q_A, Q_F, Q_V, Q_R and Q_N are positive definite diagonal weighing matrices - Q_A is again a scalar. Moreover, $g_{e_L}\left(e_{iL}(n|k), w_L(k)\right)$ is a propagation process used for prediction of the relative error with respect to the leader UAV (i.e., $e_{iL}(n+1|k)$), defined hereafter.

Leader Position Propagation:

In order to estimate the relative position of the leader along a horizon of length N, the followers measure d_{iL} and β_{iL} at the sampled instant k which, from (3.6), uniquely define $\chi_{iL}(k)$, $i \in \mathcal{F}$. Having the vector of $w_L(k)$ broadcasted from the leader to all followers, the prediction for the leader position over the receding horizon, on behalf of follower i, for $n = 0, \ldots, N-1$ is given as:

$$\begin{aligned}
\chi_{iL}(n+1|k) &= g_L\left(\chi_{iL}(n|k), w_L(k)\right) \\
&= \chi_{iL}(n|k) + R_i^T(n|k)\,[v_L(k) - v_i(k+n)]\Delta t.
\end{aligned} \tag{3.18}$$

With (3.9b), the dynamics (3.18) in error form can be rewritten

$$\begin{aligned}
e_{iL}(n+1|k) &= g_{e_L}\left(e_{iL}(n|k), w_L(k)\right) \\
&= \hat{\chi}_{iL} - \chi_{iL}(n|k) - R_i^T(n|k)\,[v_L(k) - v_i(k+n)]\Delta t \\
&= e_{iL}(n|k) - R_i^T(n|k)\,[v_L(k) - v_i(k+n)]\Delta t.
\end{aligned}$$

3.4 Stability Analysis

Before proceeding to the necessary analysis of the proposed NMPC strategy, we employ standard stability conditions that are used in MPC frameworks:

Assumption 1 *Consider the running costs* $V_{l_i}(e_i, u_i) = V_{Q_i}(e_i) + V_P(e_i, u_i)$, *and the terminal cost function* $V_V(e_i)$ *for the formation leader, such that:*

1. E_L, E_L^f, *and* U_L *are closed sets containing the origin, and* E_L^f *is a control invariant terminal set containing the origin;*

2. *The cost functions* $V_\star(e(\cdot))$, *as well as system dynamics and propagation functions* f_e *and* Υ_{e_i}, *are Lipschitz continuous, with Lipschitz constants* $L_{V_\star}$, *where* $\star = \{A, P, Q, V\}$;

Property 1 *In view of (3.9a), the difference between the nominal prediction state-error and the real state-error of the relative position of the followers relative to the leader is defined as* $\|e_{LF}(n|k) - e_{LF}(k+n)\| \leq \sum\limits_{i \in \mathcal{N}_L} \|\chi_{Li}(k+n) - \chi_{Li}(n|k)\| \leq \sum\limits_{i \in \mathcal{N}_L} \Delta v_{i_{Max}} \Delta t \cdot n.$

Regarding the followers, their prediction of the leader movement is based on the information vector $w_L(k)$. We have:

Lemma 1 *The difference between the actual state $e_{iL}(k+n)$ at time $k+n$ and the predicted state $e_{iL}(n|k)$ at the same time under the control law defined in (3.16), is upper bounded by:* $\|e_{iL}(n|k) - e_{iL}(k+n)\| \leq (1+L_f^n)\|p_L(k) - p_i(k)\| + (n+1)\cdot 2v_{i_{Max}}\Delta t$
$+ \left(\sum_{n=0}^{N} L_f^n\right) L_u\|u_L(k) - u_i(k)\|$, *where $v_{i_{Max}}$ is the maximum velocity of all $i \in \mathcal{A}$.*

Proof: Initially notice that $e_{iL}(1|k) = \hat{\chi}_{iL} - \chi_{iL}(k) - R_i^T(k)[v_L(k) - v_i(k)]\Delta t$ and $e_{iL}(k+1) = \hat{\chi}_{iL} - \chi(k+1) = \hat{\chi}_{iL} - R_i^T(k+1)[p_L(k+1) - p_i(k+1)]$. From the equations, at step $n = 1,...,N$, and the fact that $\|R_i^T(k)\| = 1, \forall k \in \mathbb{N}$ we get the general form

$$\|e_{iL}(n|k) - e_{iL}(k+n)\| =$$
$$= \| - R_i^T(k)[p_L(k) - p_i(k)] - [n+1]R_i^T(k)[v_L(k) - v_i(k+n)]\Delta t +$$
$$+ R_i^T(k+n)[p_L(k+n) - p_i(k+n)]\|$$
$$\leq \|p_L(k) - p_i(k)\| + (n+1)\cdot 2v_{i_{Max}}\Delta t +$$
$$+ \|p_L(k+n) - p_i(k+n)\| \leq$$
$$\leq (1 + L_f^n)\|p_L(k) - p_i(k)\| + (n+1)\cdot 2v_{i_{Max}}\Delta t +$$
$$+ \left(\sum_{n=0}^{N} L_f^n\right) L_u\|u_L(k) - u_i(k)\| = \gamma + \rho + \kappa,$$

where $\gamma = (1+L_f^n)\|p_L(k) - p_i(k)\|$, $\rho = (n+1)\cdot 2v_{i_{Max}}\Delta t$, and $\kappa = \left(\sum_{n=0}^{N} L_f^n\right) L_u\|u_L(k) - u_i(k)\|$. $\qquad\square$

We are now ready to state the stability theorem of this chapter:

Theorem 1 *Consider the team of UAVs described by (3.2), which are subject to constraints (3.3) and (3.4). The control inputs provided by (3.11), (3.12a)-(3.12d) as well as (3.16), (3.17a)-(3.17d) drive the errors e_L and e_i, $i \in \mathcal{F}$, to sets E_L^f and E_i^f, $i \in \mathcal{F}$, containing the origin, while satisfying the constraints imposed by the sets E_i, U_i, $i \in \mathcal{A}$, given in (3.10a)-(3.10b) and (3.4) respectively.*

Proof: Let us define $\mathcal{B}_L^n = \{z : \|z\| \leq \Delta v_{i_{Max}}\Delta t \cdot n\}, \forall n = 0\ldots N$, and $\mathcal{B}_i^n = \{z : \|z\| \leq \gamma + \rho + \kappa\}, \forall n = 0, 1, \ldots, N$. By restricting the constraints over the horizon as $E_L \ominus \mathcal{B}_L^n$, $E_i \ominus \mathcal{B}_i^n$, and employing similar arguments to [33, Theorem 1, page. 4], we conclude that for every $n = 0, 1, \ldots, N$ it holds that $u_i(n|k) \in \mathbb{U}_i$, $u_L(n|k) \in \mathbb{U}_L$, $e_i(n|k) \in E_i \ominus \mathcal{B}_i^n$, $e_L(n|k) \in E_L \ominus \mathcal{B}_L^n$, $e_i(k+N|k) \in E_i^f$ and $e_L(k+N|k) \in E_L^f$.

Define the optimal cost at step k by $J_i^*(k)$, and the feasible cost by $\bar{J}_i(k+1)$, $i \in \mathcal{A}$, and following the approach presented in [33], by employing Property-1 and

Lemma-1, we can deduce that $\Delta J_{\mathsf{L}} = \bar{J}_{\mathsf{L}}(k+1) - J_{\mathsf{L}}^*(k) \leq L_{J_{\mathsf{L}}} \sum_{i \in \mathcal{N}_{\mathsf{L}}} \Delta v_{i_{Max}} \Delta t - \|e_{\mathsf{L}}(k)\|^2$, $\Delta J_i = \bar{J}_i(k+1) - J_i^*(k) \leq L_{J_i}(\gamma + \rho + \kappa) - \|e_i(k)\|^2, \forall i \in \mathcal{F}$.

Considering the optimality of the solution, we conclude that $J_{\mathsf{L}}^*(k+1) - J_{\mathsf{L}}^*(k) \leq L_{J_{\mathsf{L}}} \Delta v_{i_{Max}} \Delta t - \|e_{\mathsf{L}}(k)\|^2$, $J_i^*(k+1) - J_i^*(k) \leq L_{J_i}(\gamma + \rho + \kappa) - \|e_i(k)\|^2$, which implies that the closed loop system is Input-to-State-Stable (ISS). This implies that the optimal costs are monotonically decreasing which consequently implies that the norm of the errors $\|e_{\mathsf{L}}(k)\|$, $\|e_i(k)\|$ converge to a neighborhood of the origin, as $k \to \infty$. $\qquad\qquad\square$

3.5 Results

In order to validate the algorithm in an application scenario, we tested a formation navigation task, composed by with 2 and 5 followers tracking a time-varying velocity.

All simulations ran on an Intel Core-i7 8750H CPU, with 16GB of RAM available, performed in a Matlab environment. The Nonlinear MPC was implemented in ACADO [34].

The desired relative positions are set as $\hat{\chi}_{\mathsf{F}_1\mathsf{L}} = \begin{bmatrix} 0 & -1 & 0 \end{bmatrix}^T [m]$, $\hat{\chi}_{\mathsf{F}_2\mathsf{L}} = \begin{bmatrix} 0 & 1 & 0 \end{bmatrix}^T [m]$. The desired attitude of the agents is kept constant and aligned with the inertial frame throughout the experiment, and therefore $\hat{\chi}_{\mathsf{LF}_i}$, for $i = 1, 2$, is implicitly defined for this test. For the second test, three followers are added to the formation, with desired relative positions $\hat{\chi}_{\mathsf{F}_3\mathsf{L}} = \begin{bmatrix} 0.65 & -0.65 & 0 \end{bmatrix}^T [m], \hat{\chi}_{\mathsf{F}_4\mathsf{L}} = \begin{bmatrix} 0.65 & 0.65 & 0 \end{bmatrix}^T [m], \hat{\chi}_{\mathsf{F}_5\mathsf{L}} = \begin{bmatrix} 1 & 0 & 0 \end{bmatrix}^T [m]$.

The first test shows the tracking capabilities of the proposed framework when a time-varying velocity is defined. This makes it possible to extend the framework to more complex scenarios. For this test, the desired formation velocity $\hat{v}_{\mathsf{L}}$ is set as

$$\hat{v}_{\mathsf{L}}(0|k+n) = \begin{bmatrix} 0.05 \\ 0.06\sin(0.5 \cdot \Delta t \cdot n + k) \\ 0 \end{bmatrix}, \forall n = 0...N - 1 \qquad (3.19)$$

leading to a sinusoidal trajectory as shown in Figure 3.3.

Figures 3.5 and 3.4 show the bearing error and velocity for each of the vehicles. Due to the time-varying trajectory, we observe that the agents achieve the equilibrium formation at $t = 10s$. For this test and for the followers, the desired velocity $\hat{v}_{\mathsf{L}}(0|k+n) = \hat{v}_{\mathsf{L}}(k)$, resulting in a bounded error for the predicted trajectory, as proved on Section 3.4. We also observe that the leader slowly accelerates to allow the formation to converge to the desired geometry at first, and then to track the desired formation velocity.

In this experiment, the average computational times were $4ms$ for the leader and $2.2ms$ for the follower. The maximum and minimum computational times are, respectively, $13ms$ and $3.5ms$ for the leader, and $7.3ms$ and $1.7ms$ for the followers. The settings for the NMPC are a receding horizon length of $N = 20$ and

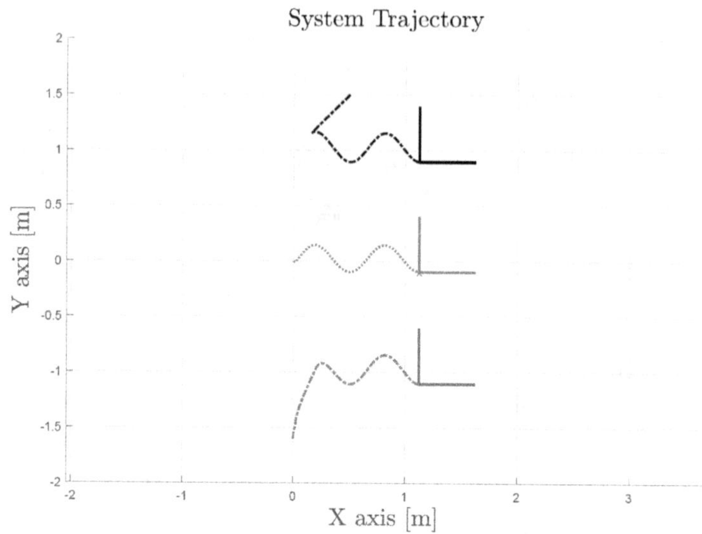

Figure 3.3: Formation trajectory in the XY inertial plane. In red is represented the leader agent, while followers F_1 and F_2 are represented in black and blue, respectively. The agent's orientation is aligned with the world frame.

a sampling time $\Delta t = 0.05s$. Finally, the maximum velocity for the agents was set to $v_{i_{Max}} = 0.1$ [m/s]. $\forall i \in \mathcal{A}$.

To better understand how the computational times are affected when followers are added to the formation, the second test was performed with 5 followers. For this test, we show the trajectory for the time-varying velocity, Figure 3.6, and the leader bearing and velocity error, on Figure 3.7.

For this test, the average computational time is $11ms$ for the leader, and $2.2ms$ for the followers. The maximum computational times are $21ms$ and $4ms$, while the minimum are $9ms$ and $1.6ms$, for the leader and followers, respectively. Although the leader's computational time increase (although not proportionally) with the number of followers, we observe that the followers controller computational time remains the same.

3.6 Discussion

In this chapter we presented a novel decentralized nonlinear model predictive control framework, with significant communication reduction and ensured stability properties. The approach is specially relevant for safety critical systems, where autonomy and resilience to failures is needed, as well as in scenarios where communication is costly, such as underwater robots. Moreover, these contributions also result in a

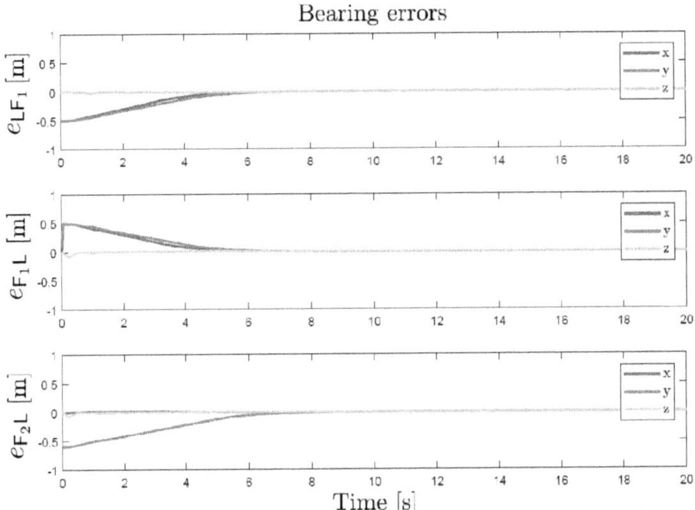

Figure 3.4: Evolution of the bearing error through time, when the formation tracks a time-varying velocity.

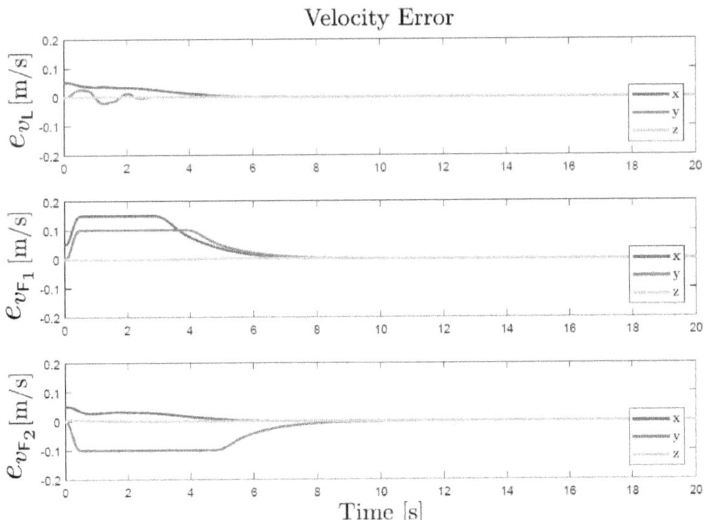

Figure 3.5: Formation velocity error over time.

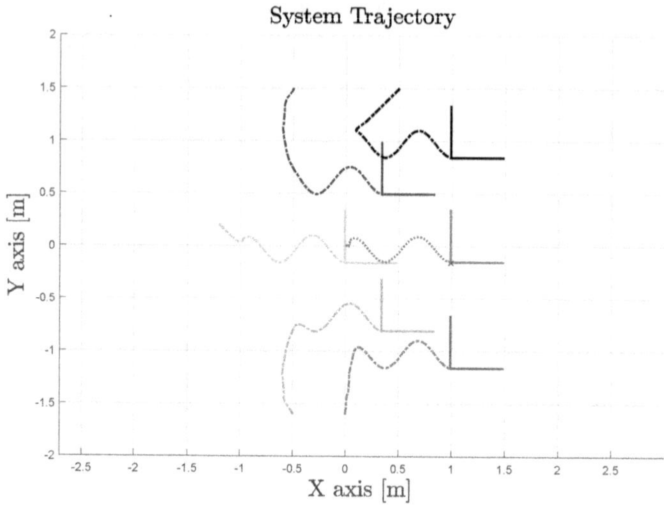

Figure 3.6: Trajectory of the formation while tracking a time-varying velocity.

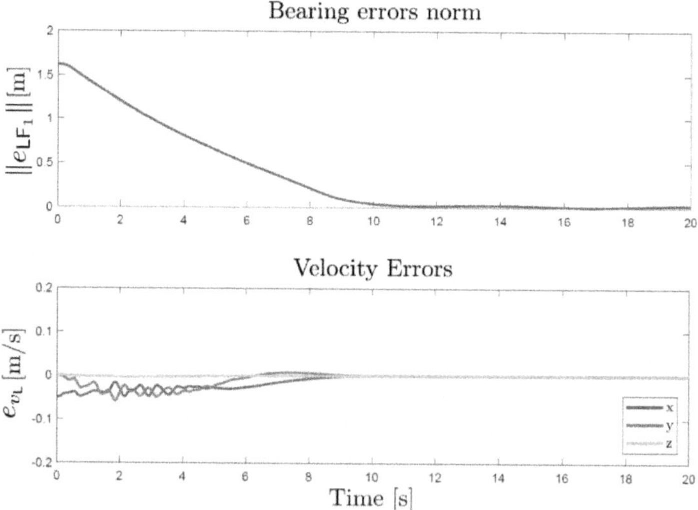

Figure 3.7: Leader bearing error norm and velocity error.

fast controller, suitable to applications with agile systems. In the future, we plan to implement these methods on a experimental scenario, as well as extend the framework to allow other graph geometries, increasing the resilience of the system and better scale the leader controller to larger formations.

Chapter 4

Formation Control for Collaborative Load Transportation

An important application of UAVs has been dealing with load transportation. Co-operating UAVs may be able to transport larger payloads that a single agent cannot, and therefore decentralized control laws are preferred for enabling efficient and robust operations. However, decentralized control laws may not always be the best choice, performance-wise. As the need for collaborative load transportation also starts to arise in orbital applications, for de-orbiting satellites or autonomous unloading of spacecrafts, it is important to explore control solutions that allow for efficient operation in such high-risk scenarios. In this chapter, we explore cooperative transportation methods and compare the performance of centralized and decentralized approaches using UAVs.

4.1   Literature Review

The maneuverability and freedom provided by UAVs is useful in a variety of applications. UAVs have already been used in search and rescue missions [35], and in inspection and defect detection of solar panels, bridges and building facades [36–38]. Another use case for UAVs is payload transport. However, the geometry and weight of the payload may often require the use of more than one UAV to make transport possible, creating a need for a controller structure with multi-agent capabilities. Furthermore, the geometric constraints introduced by the coupling between UAVs and payload, often result in complex dynamics with intricate interaction forces, rendering the control problem far from straightforward.

A common denominator for much of the previous work on UAV payload transport is to either treat the payload as a disturbance, or to find workarounds to bypass the need for modeling its complete dynamics. In [39], robust MPC with a disturbance term to account for the unknown dynamics of the load is used, and therefore cannot offer optimality with respect to its dynamics. In [40], a UAV–bar

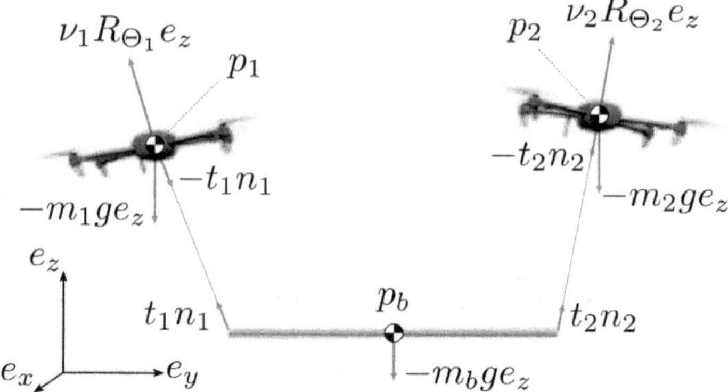

Figure 4.1: The system under analysis in this work, where two UAVs collaboratively transport a bar using cables.

system (similar to Fig. 4.1) is approximated by two UAVs with separate point-mass payloads, where each of the agents is controlled with an LQR. Other approaches take into account the full dynamics of the payload: in [2, 41] the authors design appropriate PID-based control laws to stabilize the UAV–bar system, accounting for a single equilibrium point; in [42], an offline batch optimization routine taking into account the string tension forces is used to generate appropriate trajectories. These approaches, however, lack look-ahead capabilities or online-feedback, leading to non-optimal control inputs. Lastly, in [43], open-loop control forces are designed as a function of the object's internal forces, given by end-effector sensor readings. As these are open-loop forces, steady-state error is present when no internal forces are measured.

In this chapter we propose Model Predictive Controllers (MPCs) to control the UAV–bar system depicted in Fig. 4.1. In this scenario, two UAVs collaboratively transport a cylindrical payload and two MPC control methodologies are tested: one centralized, and one decentralized. The centralized controller bears resemblance to the one found in [44], but instead of performing a numerical linearization step online, the linearization is done analytically as part of the modeling process, saving valuable computational time. The contributions are threefold: i) an analysis of the equilibrium points of the system depicted in Fig. 4.1 is provided, leading to an efficient formulation capable of being used in an online MPC framework; ii) experimental results that showcase both the centralized and decentralized control methodologies, based on the equilibrium analysis results; iii) a comparative of the results, particularly highlighting the differences between the two control setups.

4.2 Load Transportation with Aerial Vehicles

In this section, the general dynamics for load transportation using aerial vehicles are introduced, based on the state-of-the-art literature.

UAV with Single Load

Let E and B denote an inertial frame, and UAV body frame, respectively, where the frame B is attached and kept fixed to the UAV CoM as in Fig. 4.2. Let the positions $p \in \mathbb{R}^3$ and $p_p \in \mathbb{R}^3$, the velocities $v \in \mathbb{R}^3$ and $v_p \in \mathbb{R}^3$, and the masses $m \in \mathbb{R}_{>0}$ and $m_p \in \mathbb{R}_{>0}$, correspond to those of the UAV and the payload, respectively. Moreover, let g be the gravity constant, $\Theta \in \mathbb{SE}(3)$ be the Euler angles of the UAV body w.r.t. E, $M \in \mathbb{R}^{3\times3}$ its positive definite and diagonal inertia matrix and $\omega_B \in \mathbb{R}^3$ its body angular rates. Although the forces and moments that the UAV is able to generate is an effect of the combination of the thrusts generated by each propeller, a frequently used technique is to combine these thrusts into a single thrust (here $\nu \in \mathbb{R}$), and a torque vector (here $\tau \in \mathbb{R}^3$) [30, 45]. This technique is motivated by that each pair (ν, τ), uniquely defines the thrust of each propeller [45]. Then, the equations of motion of the UAV-payload system can be found using the Newton-Euler approach, resulting in

$$\dot{p} = v, \tag{4.1a}$$

$$\dot{v} = m^{-1}(\nu Re_z - mge_z - t \cdot n), \tag{4.1b}$$

$$\dot{\Theta} = H\omega^B, \tag{4.1c}$$

$$\dot{\omega}^B = M^{-1}\left(\tau - {}_\times\omega^B M\omega^B\right), \tag{4.1d}$$

$$\dot{p}_p = v_p, \tag{4.1e}$$

$$\dot{v}_p = m_p^{-1}\left(t \cdot n - m_p ge_z\right), \tag{4.1f}$$

where $n = l^{-1}(p - p_p) \in \mathbb{R}^3$ is the cable direction vector, $R \in \mathbb{SE}(3)$, abbreviated form of $R(\Theta)$, is the Z-Y-X Rotation matrix, $t \in \mathbb{R}$ is the absolute tension force in the cable attached to the payload [46, eq. (5)], H the attitude Jacobian that translates body rates to Euler angles [47], and $e_z = [0, 0, 1]^T$. These are given by

$$R(\phi, \theta, \psi) = \begin{bmatrix} c_\psi c_\theta & c_\psi s_\theta s_\phi - s_\psi c_\phi & c_\psi s_\theta c_\phi + s_\psi s_\phi \\ s_\psi c_\theta & s_\psi s_\theta s_\phi + c_\psi c_\phi & s_\psi s_\theta c_\phi - c_\psi s_\phi \\ -s_\theta & c_\theta s_\phi & c_\theta c_\phi \end{bmatrix}, \tag{4.2}$$

$$t = m_p(m_p + m)^{-1}\left(\nu Re_z \cdot n + ml^{-1}\|v - v_p\|^2\right), \tag{4.3}$$

$$H = \begin{bmatrix} 1 & s_\phi t_\theta & c_\phi t_\theta \\ 0 & c_\phi & -s_\phi \\ 0 & s_\phi/c_\theta & c_\phi/c_\theta \end{bmatrix}, \tag{4.4}$$

where the angles ϕ, θ, and ψ are referred to as roll, pitch, and yaw, respectively, and c_i and s_i represent the cosine and sine of the angle $i \in \phi, \theta, \psi$.

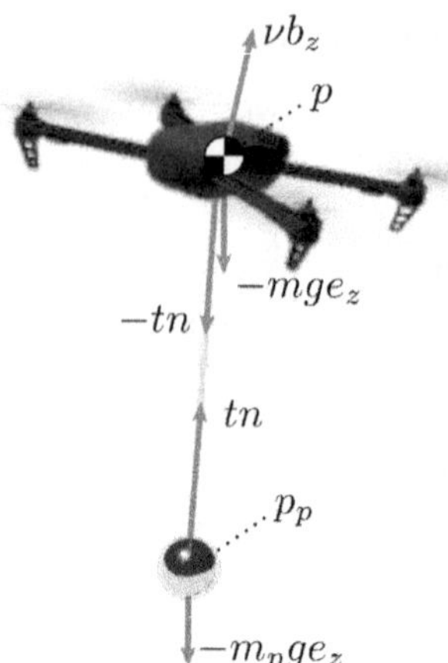

Figure 4.2: A UAV with payload attached. The body frame B has its $Z - axis$ pointing up, according to b_z. The UAV is located at a position p with respect to the inertial frame E.

UAVs with bar

We now consider a system where a rigid bar is tethered to two UAVs by cables, as illustrated in Fig. 4.1. Each UAV is given a corresponding body frame B_i, $i = 1, 2$, that is attached to the UAV CoM, and to where it is assumed that the cables are attached. The bar, however, is chosen to be modeled solely with respect to the inertial frame E. The state variables introduced in the previous section are also given subindices, so that, *e.g.*, p_1 is the position of UAV 1. The subindex b is used for quantities relating to the bar.

Each cable is of length l_1 and l_2, respectively, where the subindex is given depending on to which UAV the cable is directly attached to. We denote by m_b the mass of the bar, and d_i the distance between the bar CoM and the end of the bar to which cable i is attached. Thus, the total length of the bar is $(d_1 + d_2)$. Furthermore we define n_b as the unit vector parallel with the bar's symmetry axis, pointing towards the attachment of cable 2. Each cable direction is, in this case, given by $n_i = l_i^{-1}(p_i - (p_b + (-1)^i d_i n_b))$, $i = 1, 2$.

Using the same Newton-Euler approach as for the single UAV case results in the following set of equations

$$\dot{p}_i = v_i, \qquad\qquad i = 1, 2, b \qquad (4.5\text{a})$$

$$\dot{v}_i = m_i^{-1}\left(\nu_i R_{\Theta_i} e_z - m_i g e_z - t_i \cdot n_i\right), \qquad i = 1, 2 \qquad (4.5\text{b})$$

$$\dot{\Theta}_i = H_{\Theta_i}\omega_i^{\mathsf{B}}, \qquad\qquad i = 1, 2 \qquad (4.5\text{c})$$

$$\dot{\omega}_i^{\mathsf{B}} = M_i^{-1}\left(\tau_i - {}_\times\omega_i^{\mathsf{B}} M_i \omega_i^{\mathsf{B}}\right), \qquad i = 1, 2 \qquad (4.5\text{d})$$

$$\dot{v}_b = m_b^{-1}(t_1 n_1 + t_2 n_2 - m_b g e_z), \qquad\qquad (4.5\text{e})$$

$$\dot{\Theta}_b = H_{\Theta_b}^{\mathsf{E}}\omega_b, \qquad\qquad (4.5\text{f})$$

$$\dot{\omega}_b = M_b^{-1}{}_\times n_b(d_2 t_2 n_2 - d_1 t_1 n_1), \qquad\qquad (4.5\text{g})$$

where R_{Θ_i}, H_{Θ_i} are found by making the substitutions $\phi \quad \phi_i$, $\theta \quad \theta_i$, and $\psi \quad \psi_i$ in (4.2) and (4.4), respectively. The matrix $H_{\Theta_b}^{\mathsf{E}}$, relates angular velocities in the *inertial* frame to Euler angle rates and has the form

$$H_{\Theta_b}^{\mathsf{E}} = \begin{bmatrix} c_{\psi_b}/c_{\theta_b} & s_{\psi_b}/c_{\theta_b} & 0 \\ -s_{\psi_b} & c_{\psi_b} & 0 \\ c_{\psi_b} t_{\theta_b} & s_{\psi_b} t_{\theta_b} & 1 \end{bmatrix}. \qquad (4.6)$$

Equation (4.5g) is valid under the assumption that $n_b \cdot \omega_b = 0$, which means that the bar remains parallel to the ground does not twist in its main axis, and M_b can in this case be chosen to be the scalar corresponding to the moment of inertia in the axes perpendicular to n_b.

To reduce the model complexity, each cable is modeled as a massless, rigid, and rotational link, which means that the following constraint

$$\left\| p_i - (p_b + (-1)^i d_i n_b) \right\| = l_i, \qquad i = 1, 2 \qquad (4.7)$$

holds, and the forces $t_1(x, u)$, $t_2(x, u)$ take the form

$$\begin{bmatrix} t_1 \\ t_2 \end{bmatrix} = M^{-1}\lambda, \qquad (4.8)$$

for a matrix $M \in R^{2\times2}$, and a vector $\lambda \in R^2$, both being functions of the state and a control input $u = (\nu_i, \tau_i), i = 1, 2$. For an explicit expression, see [48].

4.3 Equilibrium points for the UAV–bar system

A result of this work corresponds to the analysis of the equilibrium points of the system depicted in Fig. 4.1.

Assumption 2 *Consider the system (4.5), where the variables t_1 and t_2 are defined according to (4.8). It is assumed that the system is symmetric with $d_1 = d_2 =: d_b$, $l_1 = l_2 =: l$, and $m_1 = m_2 =: m_{\text{uav}}$. It is also assumed that the bar is parallel with $e_x = [1, 0, 0]^T$ (inertial frame X-axis), and that it is placed with its center of mass in the origin.*

Figure 4.3: Definition of the angles θ_f and θ_l. The system is assumed symmetric with respect to the center of the bar. The variable θ_l corresponds to the angle between the cable and inertial Z-axis, while θ_f corresponds to the angle between the thrust ν_i being applied on the cable and the inertial Z-axis.

Under these conditions, it is possible to obtain a set of equilibrium points, as shown in Proposition 1.

Proposition 1 *Let Assumption 2 hold. Then the equilibrium points of* (4.5) *can be defined in closed form with respect to θ_f and θ_l, where θ_l corresponds to the angle between the cable and inertial Z-axis, and θ_f corresponds to the angle between the thrust ν_i being applied on the cable and the inertial Z-axis, according to $f_{\mathrm{eq}}(\theta_f, \theta_l) = 0$, where*

$$f_{\mathrm{eq}}(\theta_f, \theta_l) = \left(\frac{s_{\theta_f} \bar{m} g}{2 c_{\theta_f}} - \frac{m_b g c_{(\theta_f - \theta_l)} s_{\theta_l} \bar{m}}{2 c_{\theta_f} \left(2 m_{\mathrm{uav}} c_{\theta_l}{}^2 + m_b \right)} \right)^2 +$$
$$+ \left(\frac{m_b g}{2} - \frac{m_b g c_{(\theta_f - \theta_l)} c_{\theta_l} \bar{m}}{2 c_{\theta_f} \left(2 m_{\mathrm{uav}} c_{\theta_l}{}^2 + m_b \right)} \right)^2 ,$$

(4.9)

with $\bar{m} := m_b + 2 m_{\mathrm{UAV}}$.

Proof: Consider the parametrization variables θ_f and θ_l as in Fig. 4.3, the thrust forces as $\bar{\nu}_i = \nu \cdot (\sin \theta_f \cdot (-1)^i, 0, \cos \theta_f)$, $i = 1, 2$, and the string tension forces as $\bar{t}_i = t \cdot (\sin \theta_l \cdot (-1)^i, 0, \cos \theta_l)$, $i = 1, 2$. This restricts the UAV positions, so that $p_i = ((d_b + l \sin \theta_l) \cdot (-1)^i, 0, \cos \theta_l)$, $i = 1, 2$. For the vertical equilibrium of the system, we need $(\bar{\nu}_1 + \bar{\nu}_2) \cdot e_z = 2\nu \cos \theta_f = g(m_b + 2 m_{\mathrm{uav}})$, which gives

$$\nu = (2 \cos \theta_f)^{-1} \bar{m} g. \tag{4.10}$$

Similarly, for the lateral equilibrium, $(\bar{\nu}_i - \bar{t}_i) \cdot e_x = 0$, $i = 1, 2$, should hold, but the symmetry given by Assumption 2 and the definition of θ_f, θ_l implies that this can be reduced to the single equation

$$\nu \sin \theta_f - t \sin \theta_l = 0. \tag{4.11}$$

The string tension force t can be taken from (4.8) (by symmetry, $t_1 = t_2$), giving $t = \frac{m_b g \cos(\theta_f - \theta_l)(m_b + 2 m_{\mathrm{uav}})}{2 \cos(\theta_f)(2 m_{\mathrm{uav}} \cos^2(\theta_l) + m_b)}$ after substituting ν using (4.10). Finally for the bar

to be in equilibrium, it is necessary that the string tensions' vertical components
are equal to the weight of the bar, i.e.,

$$t\cos\theta_l - \frac{m_b g}{2} = 0. \tag{4.12}$$

Note that, the lateral equilibrium is fulfilled automatically due to symmetry. The
result now follows from the squared sum of (4.11) and (4.12), giving $f_{\mathrm{eq}}(\theta_f,\theta_l) = 0$.
$\square$

Out of all solutions, the trivial solution $\theta_f = \theta_l = 0$ constitutes the most efficient equilibrium point as the thrust direction is then completely aligned with the
direction of gravity. The other equilibria, however, remain interesting in situations
where the space in the z-direction is limited, and also enables the possibility of rais-
ing the bar while the z-position of the UAVs remains relatively intact (or lowering
the UAVs while the bar remains in position.)

4.4 Centralized Control

This controller makes use of the model derived in Section 4.2. To reduce the system
complexity and consequently the MPC solving time we make use of a linearized
version of (4.5), leaving the yaw of each rigid body as a parameter that is set at each
sampling time. As linearization point, any of the equilibria defined by the solution
to (4.9) can be chosen, as long as Assumption 2 holds. While the bar can assume an
arbitrary position p_b, the positions of the UAVs are decided explicitly based on the
yaw and the position of the bar if the system is to remain in the chosen equilibrium
position. Consequently, the equilibrium state is given by $x_{\mathrm{eq}} = (x_{\mathrm{eq}}^{(b)}, x_{\mathrm{eq}}^{(1)}, x_{\mathrm{eq}}^{(2)})$,
where $x_{\mathrm{eq}}^{(b)} = (p_b, 0_5, \psi_0^{(b)}, 0_3)$, $x_{\mathrm{eq}}^{(i)} = (p_i, 0_4, (-1)^{i-1}\theta_f, \psi_0^{(i)}, 0_3)$, $i = 1, 2$, and $p_i =$
$p_b + ((-1)^{i-1}(d_i + l_i \sin\theta_l)\sin\psi_0^{(b)}, (-1)^i(d_i + l_i \sin\theta_l)\cos\psi_0^{(b)}, l_i \cos\theta_l)$, $i = 1, 2$. The
corresponding input equilibrium is given by the conditions of force and moment
equilibria which result in $u_{\mathrm{eq}} = (\nu_{\mathrm{eq}}^{(1)}, \tau_{\mathrm{eq}}^{(1)}, \nu_{\mathrm{eq}}^{(2)}, \tau_{\mathrm{eq}}^{(2)})$, where $\tau_{\mathrm{eq}}^{(1)} = \tau_{\mathrm{eq}}^{(2)} = 0_3$,
and $\nu_{\mathrm{eq}}^{(i)} = g\bar{m}/(2c_{\theta_f})$. The resulting linearized system equations can be written
as $\dot{x}_{\mathrm{lin}}(t) := A_c\Delta x(t) + B_c\Delta u(t)$, where $A_c = \partial\dot{x}/\partial x\big|_{x=x_{\mathrm{eq}}}$, $B_c = \partial\dot{x}/\partial u\big|_{x=x_{\mathrm{eq}}}$,
$\Delta x := x - x_{\mathrm{eq}}$, and $\Delta u := u - u_{\mathrm{eq}}$. The standard ZOH discretization of $\dot{x}_{\mathrm{lin}}$ would
result in many higher order terms of the parameters $\psi_0^{(b)}$, $\psi_0^{(1)}$, and $\psi_0^{(2)}$, and we
therefore use a first order approximation $x((k+1)\Delta t) = A\Delta x(k\Delta t) + B\Delta u(k\Delta t)$,
where $A = I + A_c\Delta t$, $B = B_c\Delta t$, and Δt is the sampling time.

Finally the MPC optimization problem is given by

$$\min_{\mathbf{u}_k^*} \sum_{n=0}^{N-1} \left(\|\Delta x(n\Delta t | k\Delta t)\|_Q^2 + \|\Delta u(n\Delta t | k\Delta t)\|_{R_{\mathrm{MPC}}}^2 \right)$$
$$+ \|\Delta x(N | k\Delta t)\|_{Q_N}^2$$

$$\text{s.t.} \quad x((m+1)\Delta t | k\Delta t) = Ax(m\Delta t | k\Delta t) + Bu(m\Delta t | k\Delta t), \qquad \text{(MPC)}$$
$$x(m\Delta t | k\Delta t) \in \mathbb{X},$$
$$u(m\Delta t | k\Delta t) \in \mathbb{U}, \qquad \forall m \in \mathbb{N}_{[0, N-1]},$$
$$x(N\Delta t | k\Delta t) \in \mathbb{X}_N,$$
$$x(0 | k\Delta t) := x(k\Delta t)$$

for any $k \in \mathbb{N}_0$, and where the full state vector x is ordered

$$x = [p_b, v_b, \Theta_b, \omega_b, p_1, v_1, \Theta_1, \omega_1, p_2, v_2, \Theta_2, \omega_2]^T \in R^{36}$$

the matrices Q, Q_N, R_{MPC} are user-designed positive definite weighting matrices, and the constraint sets $\mathbb{X}$, $\mathbb{X}_N$, $\mathbb{U}$ define the state, terminal state and control input constraints. The solution to (MPC) will be a set of vectors $\mathbf{u}_k^*$, out of which only the first, $u(0 | k\Delta t) =: u^*$, will be applied.

4.5 Decentralized Control

In this section, we aim to decentralize the controller structure defined in the previous section. While there could be many benefits of a decentralization, the main point of interest within this work is to investigate how this decentralization affects the MPC solving time and the overall performance of the closed-loop system.

Since the UAVs are dynamically coupled by the cable tension force T, the original system cannot be decentralized without modifications. For this reason, the UAV–bar system is approximated further by instead considering it as two independent UAVs with payloads. Hence, the endpoints of the bar are tracked, and their position and velocity considered as the positions and velocities of imaginary payloads attached to each UAV by cables. The weight given to each (imaginary) payload is set to be the same weight that the bar would apply to the cable in the U-shaped equilibrium pose where $n_1 = n_2 = e_z$ and $n_b \cdot e_z = 0$, which would be equal to $m_b/2$ in a symmetric setup, but more generally

$$m_p^{(i)} := m_b d_i^{-1} (d_1 + d_2)^{-1} d_2 d_1, \qquad (4.13)$$

Using the model description in Section 4.2, the model used in the controller of UAV i is then found by updating m_p using (4.13) in (4.1f) and (4.3), followed by a linearization and discretization as explained in Section 4.4. Once the models are derived, an MPC can be designed for each UAV as in (MPC). In other words, each UAV (with payload) defines its own MPC optimization problem, each of which can

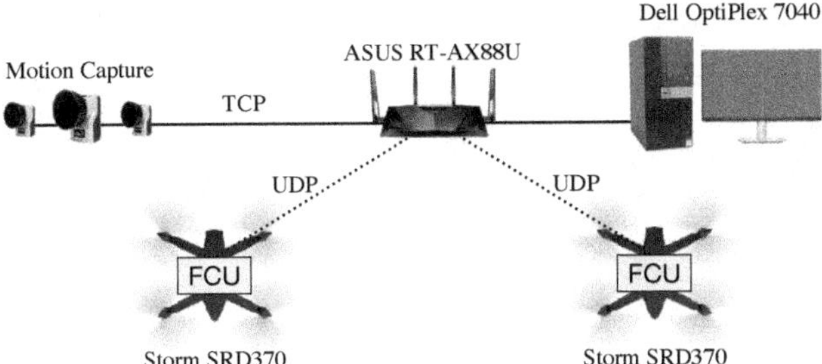

Figure 4.4: An illustration of the experimental setup where solid (dashed) lines represent wired (wireless) connections.

be solved independently, making this a decentralized setup. We have chosen the ordering of the state vector of UAV i as $x_i = [p_i, v_i, \Theta_i, \omega_i, p_p, v_p] \in R^{18}$.

4.6 Experimental Setup

Two Storm SRD370 quadrotors equipped with onboard mRo Pixracer FCUs flashed with PX4 firmware version `1.10.1`, were connected with a host computer (Intel i7-6700@3.4GHz, 8 threads; 32 GB DDR4@2133 MHz) over Wi-Fi and communicated using the `MAVLink` (`https://mavlink.io`) messaging protocol. The host computer used `MAVLink` accompanied with `MAVROS`, which enabled the translation of ROS messages to the relevant `MAVLink` messages. A motion capture system (Qualisys) was used for receiving state estimates. Fig. 4.4 provides a schematic of the network.

Both MPCs were implemented as a ROS node and the optimal control problems formulated in `CasADi` [49] and solved using the `HPMPC` solver. The parameters of the centralized MPC were set as seen in Table 4.1, and the decentralized MPC as in Table 4.2 for both UAVs. For both controllers the input constraints $\mathbb{U}$ were set in order to guarantee a positive thrust ($\nu_i > 0$) and such that the torque magnitudes were less than or equal to one ($\|\tau_i\|_\infty \leq 1\,\mathrm{N\,m}$). Initial benchmarks of the CPU time required by the MPCs, gave results averaging sligthly below $10\,\mathrm{ms}$, and thus the sampling time Δt was set to $20\,\mathrm{ms}$ to allow for occasional CPU time increases.

The experiments were performed at the $6 \times 6 \times 3\,\mathrm{m}^3$ arena at the KTH Smart Mobility Lab, where both controllers were setup to track four setpoints in sequence. As payload, a hollow bar weighing $0.39\,\mathrm{kg}$ and of length $1.47\,\mathrm{m}$ was used, and the Storm SRD370s were modeled as having masses $m_i = 1.15\,\mathrm{kg}$, and moments of inertia $M_i = \mathrm{diag}(0.0348, 0.0459, 0.0977)[kgm^2]$. During initial tests of the decentralized MPC setup it was discovered that the assumption that the bar was homogenous with a CoM in the center of its extension along the axis of symmetry was not accurate enough. Instead, d_1 and d_2 were re-identified and found to be $d_1 = 70.5\,\mathrm{cm}$ and $d_2 = 76.5\,\mathrm{cm}$, and the values of $m_p^{(i)}$ were updated according to

Table 4.1: Parameter settings the centralized MPC.

Q	$\mathrm{diag}(Q'_b, Q'_{\mathrm{UAVs}}, Q'_{\mathrm{UAVs}})$
Q'_b	$\mathrm{diag}(50, 50, 50, 50, 50, 50, 10, 10, 10, 1, 1, 5)$
Q'_{UAVs}	$\mathrm{diag}(30, 30, 30, 30, 30, 30, 10, 10, 20, 10, 10, 100)$
R_{MPC}	$\mathrm{diag}(1, 100, 100, 50, 1, 100, 100, 50)$
Q_N	$100Q$
Δt	$0.02\,\mathrm{s}$
N	30

Table 4.2: Parameter settings for the decentralized MPCs.

Q	$\mathrm{diag}(Q'_{\mathrm{UAV}}, Q'_p)$
Q'_{UAVs}	$\mathrm{diag}(20, 20, 20, 30, 30, 30, 10, 10, 30, 10, 10, 30)$
Q'_p	$\mathrm{diag}(20, 20, 20, 1, 1, 1)$
R_{MPC}	$\mathrm{diag}(1, 100, 100, 100)$
Q_N	$50Q$
Δt	$0.02\,\mathrm{s}$
N	30

(4.13).

4.7 Results

In this section we analyze the results obtained by implementing the controllers in Sections 4.4 and 4.5 to a real system. An accompanying video showcases the experiments. Before setting up the experimental platform, both controllers were tested in a Gazebo simulation environment running the PX4 software in the loop (SITL) using the following setpoints (Fig. 4.5): first the bar is rotated 90 degrees around its centerpoint, while UAV 1 also rotates an equal amount. Then the bar is translated a distance of one meter while both UAVs rotate 90 degrees. After returning back to the original position, the system is then set to the equilibrium point described by $\theta_f = 0.1183$, $\theta_l = \pi/4$ (this is a solution to (4.9) for $m_{\mathrm{UAV}} = 1.52$ kg, $m_b = 0.41$ kg, and $g = 9.81\,\mathrm{m/s^2}$) before finally assuming the initial configuration.

Centralized MPC

Figures 4.6a and 4.6b show the position and orientation of all rigid bodies, as well as the reference signals (in dashed lines). Figures 4.7a and 4.7b show the resulting optimal control signals. The overall behavior of the controller is consistent with initial simulation results (left out for brevity): the system is able to track all setpoints, both in terms of position and orientation. In Fig. 4.7a it can be noted

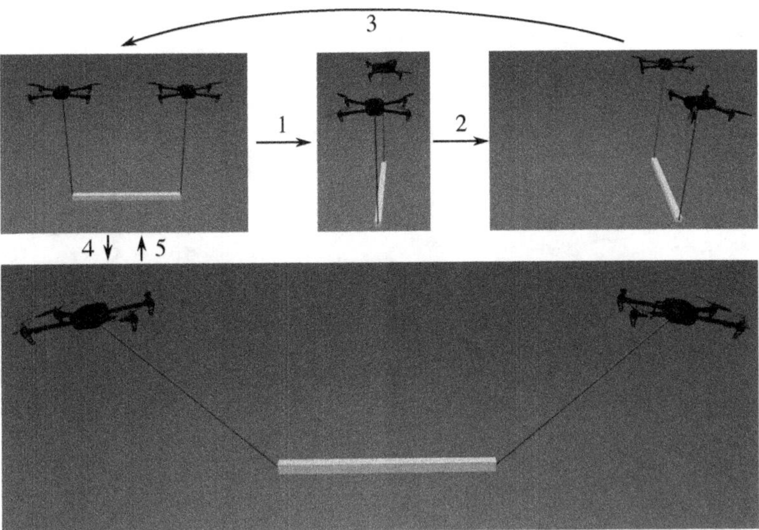

Figure 4.5: Images of the setpoints used in a preliminary simulation using the PX4
SITL.

that the thrust of UAV 1 seems to be of consistently larger magnitude than that of
UAV 2, which is inline with the non-symmetric placement of the bars CoM discussed
in Section 4.6.

Decentralized MPC

Figure 4.6c shows position of all rigid bodies, as well as the reference signals (in
dashed lines); however, as the bar centerpoint was no longer being tracked, its
estimated position was calculated as $p_b = p_p^{(1)} + 2^{-1}(d_1 + d_2)(p_p^{(2)} - p_p^{(1)})$. Fig. 4.6d
shows the orientations of the UAVs, as well as the estimated yaw angle of the
bar, calculated by $\arctan2(y_p^{(2)} - y_p^{(1)}, x_p^{(2)} - x_p^{(1)})$, where $x_p^{(i)}$, $y_p^{(i)}$ are the x and y
positions of payload i, and $\arctan2 : \mathbb{R}^2 \to (-\pi, \pi]$ is the function commonly found
in many programming languages which, in contrast to arctan, returns values in the
range $(-\pi, \pi]$. Figures 4.7c and 4.7d show the resulting optimal control signals,
where it should be noted that the relatively large difference between the thrust
signals of UAV 1 and 2 likely stems from the differences in $m_p^{(i)}$ (eq. (4.13)) as
discussed in Section 4.6.

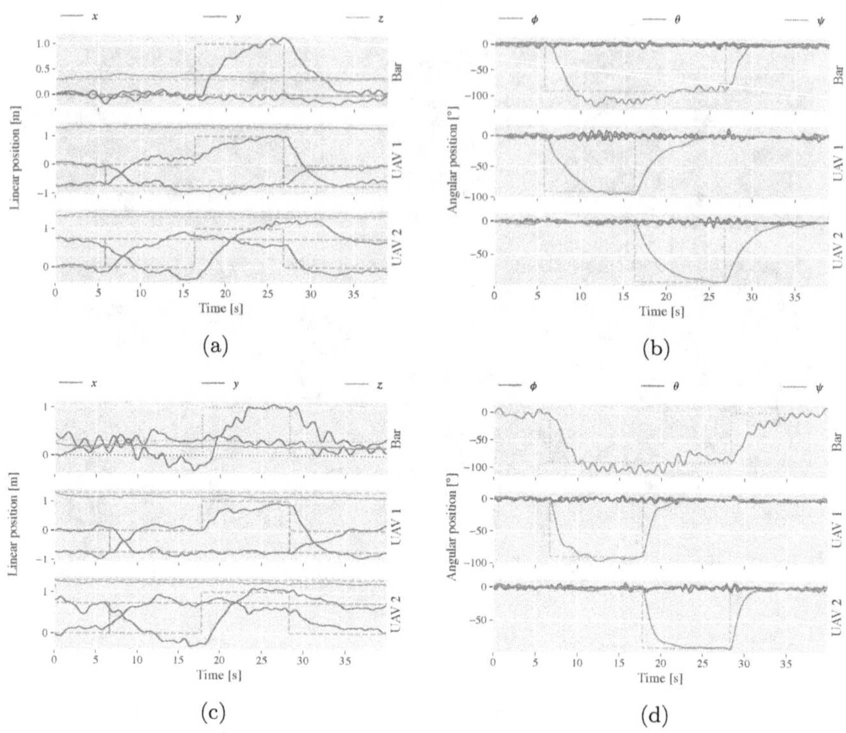

Figure 4.6: Experimental results showing position and orientation of both UAVs and bar for the: (a), (b) centralized MPC; (c), (d) decentralized MPC. Dashed lines denote the reference signal.

Comparison

Figure 4.8 shows errors and magnitudes that will aid in the comparison: Fig. 4.8a the L^2-norm of UAV 1's error in position (top) and orientation (bottom); Fig. 4.8b the thrust magnitude (top) and the L^2-norm of the torques; Fig. 4.8c the absolute error in all three dimensions of the bar centerpoint position (top) and the error in yaw orientation (bottom). Starting with Fig. 4.8a, we notice that the position errors of the UAV are close to identical. In the case of the orientation, the decentralized setup achieves a faster convergence time, which is likely a result of the difference in tuning. The results in Fig. 4.8b do not indicate any clear differences between the two setups, apart from the temporary increase in torque magnitude for the centralized setup occurring between sec. 10–15. Since this event coincides with UAV 1 being rotated $-90°$, this could eventually be caused by a failure to attach the cables to the UAV CoM, thereby introducing rotational forces that are not part

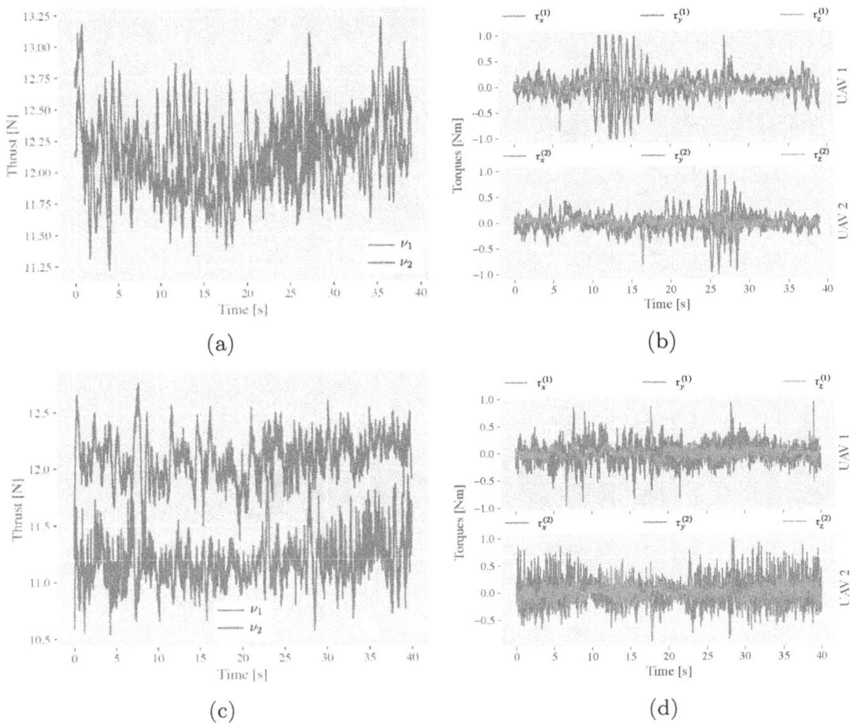

Figure 4.7: Experimental results showing thrust and torques of both UAVs for the: (a), (b) centralized MPC; (c), (d) decentralized MPC.

of the system model. The noise-affected appearance of the input signals could be attributed to that: i) the velocity estimate is noisy; ii) the transfer of the input from the MPC to the vehicle is sub-optimal, resulting in overshoots that the MPC has to compensate for. Figure 4.8c, in contrast, displays a notable difference between the two setups: in the decentralized (dashed) case, the position and yaw angle error seem to have a more oscillatory character, and the yaw error convergence between sec. 27–35, is slower. To quantify this we use the error in the x-position as a proxy, and calculate the arithmetic mean and sample standard deviation for the two cases. Note that perfect tracking would be equivalent to both these quantities being zero. Results are given in Table 4.3 and give the decentralized MPC a mean approximately four times further away from zero, and a two times higher standard deviation compared to the centralized MPC. Another important comparison factor is the CPU time: during the experiments a time average of 2.8 ms, and 7.0 ms is achieved for the decentralized and centralized case, respectively. The decentralized MPC was thus found to have a 2.5 times faster execution time.

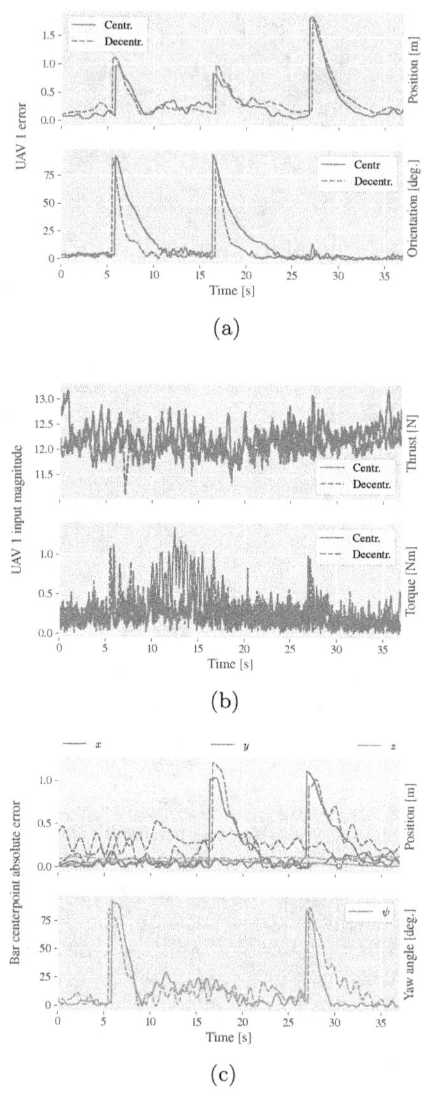

(a)

(b)

(c)

Figure 4.8: Plots showing the experimental results in terms of: (a) UAV 1 error; (b) input magnitude; (c) bar error. Dashed lines denote results from the decentralized setup.

Table 4.3: Statistics of the error in bar centerpoint x-position.

	Mean	Std. dev.
Centralized MPC	$-0.055\,14$ m	$0.058\,46$ m
Decentralized MPC	$0.234\,68$ m	$0.125\,54$ m

A conclusion can then be drawn: a centralized approach where the dynamics of the bar are incorporated results in better tracking of the bar, but comes at the cost of increased computational time and lack of redundancy. However, there are also characteristics, inherent to each setup, that should be considered in the choice between a centralized and a decentralized setup. By decentralizing the control problem, a higher resilience towards system failure is achieved, since the failure of one agent does not necessarily mean the failure of the rest. Possibly, the closed-loop performance of the decentralized setup could be improved by incorporating more feedback into the model. As an example, the relative distance between UAVs could be tracked and fed into the controller.

4.8 Discussion

The aim of this work was to investigate the effectiveness and performance of model predictive controllers when applied to the case of collaborative payload transport using UAVs. For this purpose, two different MPCs were proposed and designed: a centralized, and a decentralized. Both controllers were set up on an experimental platform where they were able to track a number of setpoints successfully. The centralized MPC gave better tracking of the bar, but had a higher CPU time than the decentralized. Both controllers seemed to perform equally in terms of UAV tracking.

This work demonstrated that MPC can be utilized to control a UAV–bar system. Future work includes the consideration of imperfect state estimation, as well as systems with more UAVs and more complicated payload geometries. Although a decentralized attempt was done in this work, the performance could likely be improved by further investigation into *e.g.*, if there are any quantities that should be transmitted between agents, and/or looking at if it is possible to incorporate more of the of the dynamics caused by the cable tension without adding to the model complexity.

Chapter 5

Vision-based Predictive Control for Setpoint Stabilization

Vision sensors are being widely used in robotics due to their low-cost and high-reward for understanding the surrounding environment. Control laws that can take advantage of these sensors are therefore desireable, both to improve the performance of vision-aware algorithms and for reliable navigation based on proprioception. In this chapter, we explore the usage of a Linear MPC for fast control updates and an IBVS module, with the goal of setpoint stabilization. We provide guarantees on the stability of the proposed Linear MPC, and both simulation and experimental results with a quadcopter platform.

5.1 Literature Review

Guided by its applications, *e.g.*, search and rescue [50], inspection [51, 52], mapping and exploration [53], photography [54], agriculture [55], and construction [56], research on UAVs has been one of the more active topics in robotics. There are many types of UAVs that can be categorized depending on their dimension and number/type of actuators. The two main categories are fixed-wing *v.s.* rotary-wing UAVs. In this work we focus on the latter, namely the use of quadrotors. These vehicles are characterized by having four rotational propellers. Quadrotor UAVs are underactuated [30]. The reason for this is the fact that there is no direct actuation providing motion on the xy–direction. In this chapter, we tackle the control of a quadrotor using computer vision, namely Visual Servoing (VS) [18, 57]. VS aims at providing control inputs for a camera (attached to a robotic agent) to take it from an initial to a goal position in the environment. However, the output of the VS module provides six degrees of freedom control inputs (linear and angular velocities) to the camera, which is not suitable for the quadrotor's under-actuated dynamics.

This chapter tackles the under-actuation issues of the quadrotor VS using MPC

[58]. In [59] a preliminary work to navigate a UAV with IBVS is presented. We model a linear MPC scheme with the quadrotor dynamics and the VS control velocity goal, providing the actuation inputs to the UAV agent, *i.e.*, roll, pitch, yaw, and thrust quantities. Different from existing works, such as [60], we focus on the approximation using a linear MPC, with the purpose of having a fast, low cost solver. To the best of our knowledge, this is the first time where VS is successfully combined with MPC for low power quadrotors, enabling its application in real-time scenarios. The conditions for the feasibility and stability of the linear MPC are studied, and the proposed control scheme is tested in simulated and real scenarios.

In addition to fixing the under-actuation issues of quadrotors, with the proposed framework we can efficiently include additional restrictions on the motion of the robotic agent. An advantage of the proposed formulation with respect to the state-of-the-art is the possibility to include a system model to track velocities that non-holonomic systems cannot, in general, track optimally. Moreover, the inclusion of state and control constraints means that this framework can handle actuation limits from the hardware. Also worth noting, it is possible to use previous control predictions to actuate the system, in the event of a feature detection failure.

Despite the related background, the coupling of VS with MPC to model the UAV's dynamics has not yet fully studied. In [61, 62], the authors propose the use of MPC for aircraft collision avoidance and use single point features to guide the robotic agent around the object, along a conical spiral trajectory. [63] proposes an autonomous vision-based landing of helicopter-based unmanned aerial vehicle (UAV). A real-time MPC-based optimization scheme to drive the robotic agent through the requested flight pattern is proposed. In [60], the authors present an observer-based model predictive control scheme for quadrotors to explicitly bound the roll and pitch angles and alleviate the feature loss on large rotation.

Other relevant works on similar topics are available in the literature. [64, 65] concerns IBVS for aerial vehicles. The authors use a PID to track a desired thrust and angular momentum's instead of MPC, and no actuation limits are considered. [66] also did not account for actuation limits and focuses on planning and low-level feedback control. [67–69] aim at a high level planning, instead of low-level control. The high computational cost involved required a powerful off-board computer. In [70], the authors propose a nonlinear MPC with visual servoing (focusing the center of the image on a given set of points). However, no guarantees of stability are given, which we do due to our linear MPC nature.

In addition to the use of MPC schemes for the control of aircrafts using computer vision, MPC was used in several works to help with the issues of the IBVS, namely the full-actuation assumption and to deal with keeping image features within the image frame. In [71, 72], the authors propose control schemes for IBVS for manipulators with eye-in-hand configurations. [73] studies a similar problem, in which the authors include the possibility of having a wide field of view camera in the manipulator's hand effector. In [74], the authors use an MPC to simultaneously solve the problem of feature correspondence and provide control inputs, and in [75] a Nonlinear MPC is used to guide fixed-wing aircraft around an obstacle. To the best of our knowledge, there is no work that studies the use of MPC to deal with the under-actuation issues raised in the VS control loop in quadrotors.

5.2 Problem Statement

Let us consider the normalized image feature error in (2.21), given by

$$\tilde{f}_i = f_i - \bar{f}_i \;\Rightarrow\; \dot{\tilde{f}}_i = \dot{f}_i. \tag{5.1}$$

and the interaction matrix in (2.24) for perspective cameras, given by

$$L_i = \begin{bmatrix} -\frac{1}{z_i} & 0 & \frac{u_i}{z_i} & u_i \cdot v_i & -(1+u_i^2) & v_i \\ 0 & -\frac{1}{z_i} & \frac{v_i}{z_i} & 1+v_i^2 & -u_i \cdot v_i & -u_i \end{bmatrix}. \tag{5.2}$$

By inverting equation (5.1) and imposing an exponential error decrease profile (*i.e.*, $\dot{\tilde{f}}_i = -\eta \tilde{f}_i$ where η is a control gain), we obtain the velocity control law

$$\gamma_{\mathsf{c}} = -\eta L_i^+ \tilde{f}_i, \tag{5.3}$$

where $L_i^+ \in \mathbb{R}^{6 \times 2j}$ is the Moore-Penrose pseudo-inverse of L_i and j is the number of features being tracked, and $\gamma_{\mathsf{c}} = [v_{\mathsf{c}}^T \omega_{\mathsf{c}}^T]^T$ is composed by a linear velocity component $v \in \mathbb{R}^3$ and an angular velocity component $\omega \in \mathbb{R}^3$, both in the camera frame.

IBVS Stability: The stability of the IBVS control law in (5.3) is dependent on the rank of the interaction matrix, which needs to be full rank to avoid singularities. For this condition to be met, the system needs to track at least 3 features. Tracking 3 features, however, may still lead to some configurations in which ηL_i^+ is singular. Moreover, with only 3 points, there exist four local minima poses [57] (while only one will be correct, up to four valid solutions can be computed for which $\tilde{f}_i$ converges). To avoid this, we track 4 non-collinear features.

System Model

The state of the quadrotor is defined by its velocity v in the inertial frame, its attitude $\alpha := [\theta, \phi, \psi]$ with respect to the inertial frame, where θ, ϕ and ψ are the rotation with respect to the inertial x–axis (roll), y–axis (pitch) and z–axis (yaw), and its angular velocity on the body frame ω, respectively. These states are concatenated under

$$\begin{aligned} x \in \mathbb{R}^9 &:= \begin{bmatrix} v & \alpha & \omega \end{bmatrix} \\ &= \begin{bmatrix} v_x & v_y & v_z & \theta & \phi & \psi & \omega_x & \omega_y & \omega_z \end{bmatrix}, \end{aligned} \tag{5.4}$$

Figure 5.1: Quadrotor model. Axis x, y and z represent the vehicle body frame axis. Constant a_l represents the arm length of the motors, corresponding to the distance of the motors to the origin of the body frame.

where the subscripts indicate the axis of reference. The quadrotor non-linear dynamics is defined as in [30]:

$$\dot{v} = R_{\mathsf{B}}\frac{\nu}{m} + g, \tag{5.5a}$$

$$\dot{\alpha} = T\omega, \text{ and} \tag{5.5b}$$

$$\dot{\omega} = M^{-1}(\tau - \omega \times M\omega), \tag{5.5c}$$

where m and $M \in \mathbb{R}^{3\times 3}$ are the mass and inertia matrix of the UAV, $T \in \mathbb{R}^{3\times 3}$ is the body to the inertial frame attitude Jacobian, and $R_{\mathsf{B}} \in \mathcal{SO}(3)$ the rotation matrix from the body to inertial-frame. In particular, the inertia is a diagonal matrix with components $M_{[x]}, M_{[y]}$ and $M_{[z]}$ for each axis. Vectors ν and τ are respectively the force and torque vectors, where the force vector only has a body z-axis component, due to the geometrical location of the motors in the UAV body – see Fig. 5.1:

$$\nu = \begin{bmatrix} 0 & 0 & \nu_z \end{bmatrix}. \tag{5.6}$$

Due to this property, the dynamics in equations (5.5) are under-actuated.

The set where all states x lie in is denoted as X. The set of control inputs ν and τ is denoted as U, defined as

$$U \triangleq \{-\frac{\nu_z}{4}a_l \leq \tau_{x,y} \leq \frac{\nu_z}{4}a_l, 0 \leq \nu_z \leq 4 \cdot 9.81 \cdot m\}, \tag{5.7}$$

where a_l is the distance from the center of the UAV to the center of the rotors. As ν_z is the total thrust produced by the 4 rotors, the torque that the vehicle can apply is the force produced by one rotor, multiplied by the lever-arm a_l. Finally, the vehicle cannot produce an acceleration more than 4 times its mass (based off the available motors for the experimental setup).

For IBVS problems, the movement of the camera is assumed to be small. Therefore, the the vehicle will operate close to the hovering equilibrium point, around which we linearize the dynamics in (5.5), *i.e.*,

$$x_{eq} = 0, \tag{5.8}$$

which can be written in the compact form as

$$\dot{x} = f_c(x, u) = A_c x + B_c u, \tag{5.9}$$

with

$$A_c \in \mathbb{R}^{9 \times 9} = \begin{bmatrix} 0^{3 \times 3} & [g]_\times & 0^{3 \times 3} \\ 0^{3 \times 3} & 0^{3 \times 3} & I^{3 \times 3} \\ 0^{3 \times 3} & 0^{3 \times 3} & 0^{3 \times 3} \end{bmatrix}, \quad [g]_\times := \begin{bmatrix} 0 & -g & 0 \\ g & 0 & 0 \\ 0 & 0 & 0 \end{bmatrix}, \tag{5.10}$$

$$B_c \in \mathbb{R}^{9 \times 4} = \begin{bmatrix} 0^{3 \times 3} & \frac{1}{m}\epsilon_z \\ 0^{3 \times 3} & 0^{3 \times 1} \\ M^{-1} & 0^{3 \times 1} \end{bmatrix}, \tag{5.11}$$

where

$$\epsilon_z = \begin{bmatrix} 0 & 0 & 1 \end{bmatrix}^T \tag{5.12}$$

and $I^{3 \times 3} \in \mathbb{R}^{3 \times 3}$ are the 3×3 identity matrix. The control input vector u for the linearized system is then composed by

$$u = \begin{bmatrix} \tau^T & \nu_z \end{bmatrix}^T. \tag{5.13}$$

Finally, we consider a discretization of the continuous model, using a sampling time Δt. In this case, we obtain the linearized discrete-time model

$$x(k+1) = f(x(k), u(k)) = Ax(k) + Bu(k), \tag{5.14}$$

where A and B are the discrete-time state-space matrices for A_c and B_c, through

$$A = e^{A_c \Delta t} = I^{9 \times 9} + \Delta t \begin{bmatrix} 0^{3 \times 3} & [g]_\times & [r]_\times \\ 0^{3 \times 3} & 0^{3 \times 3} & I^{3 \times 3} \\ 0^{3 \times 3} & 0^{3 \times 3} & 0^{3 \times 3} \end{bmatrix}, \quad \text{where}$$

$$r = \begin{bmatrix} 0 \\ 0 \\ 0.0491 \end{bmatrix}, \quad \text{and} \tag{5.15}$$

$$B = \int_0^h e^{A_c v} B_c dv = \Delta t \begin{bmatrix} H & \frac{1}{m}\epsilon_z \\ \frac{1}{2}M^{-1} & 0^{3 \times 1} \\ M^{-1} & 0^{3 \times 1} \end{bmatrix}, \quad \text{where}$$

$$H = \begin{bmatrix} 0 & -\frac{0.016}{M_{[y]}} & 0 \\ \frac{0.016}{M_{[x]}} & 0 & 0 \\ 0 & 0 & 0 \end{bmatrix}, \tag{5.16}$$

noting that $[g]_\times$ and $[r]_\times$ corresponds to the skew-symmetric matrix of g and r, respectively.

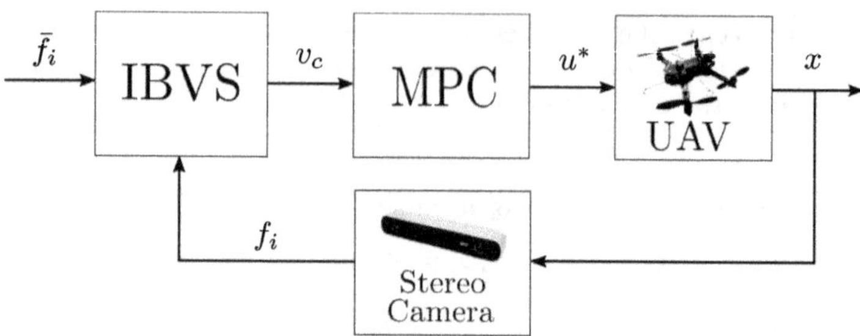

Figure 5.2: System architecture: at each camera sampling time, we detect and match the features of interest. IBVS provides a desired velocity in the camera frame, which is transformed to the UAV body. This velocity is then used as a reference to be tracked by the MPC framework, yielding optimal control inputs that are then sent to the hardware-level controller (PX4 [76]). Note that the Stereo Camera is attached the UAV frame.

5.3 Model Predictive Image-based Visual Servoing for Quadrotors

In this section we present the proposed Image-Based Visual Servo control (IBVS) for quadrotors using Model Predictive Control (MPC). To combine both modules, we implemented the cascaded control system, as shown in Fig. 5.2. We start by modeling the IBVS with MPC. Then, we analyze its stability.

Model Predictive Control

The MPC module is responsible to track the desired velocity given by the IBVS module, considering the under-actuated constraints of the quadrotor.

At each step $n = 0...N-1$ where N is the horizon length of the MPC formulation, and from an initial error $x^*(t)$, an optimal control input $u(t|t)$ is obtained by minimizing a cost function $J(\cdot)$, yielding a predicted state $x^*(1|k)$. Recursively performing these steps, yields a set of optimal predicted states $\{x^*(1|k), x^*(2|k), ..., x^*(t + N|t)\}$ and $N-1$ optimal control inputs $\{u(0|k), u(1|k), ..., u(k+N-1|k)\}, \forall k \in \mathbb{N}_0$, and where $k\Delta t$ is abbreviated by k for simplicity. Finally, at each sampling time of the MPC controller, the first optimal control input $u(0|k)$ is selected to be applied to the system.

The desired state for system (5.5) is

$$\bar{x} = \begin{bmatrix} R_\mathsf{B} R_{\mathsf{c}/\mathsf{B}} v_\mathsf{c} \\ 0^{3 \times 1} \\ \omega_\mathsf{B} \end{bmatrix}, \tag{5.17}$$

with

$$\omega_{\mathsf{B}} = \begin{bmatrix} 1 & 0 & 0 \\ 0 & 1 & 0 \\ 0 & 0 & 0 \end{bmatrix} R_{\mathsf{c/B}}\omega_{\mathsf{c}}, \tag{5.18}$$

where v_{c} and ω_{c} are given in (5.3), and where $R_{\mathsf{c/B}}$ and R_{B} represent the camera to UAV body and UAV body to inertial rotation matrices, respectively.

With this choice of $\bar{x}$, we close the loop of the IBVS module, and we use MPC to track the received velocity and cope with the system dynamics to obtain an optimal tracking controller. Note that we disregard the angular velocity around the body z-axis - this is due to the selected system linearization state, which considers a fixed heading, as at this point we are interested on translating the UAV to desired visual navigation marks.

In this subsection we define and analyze the proposed MPC framework. First, let the error state $\tilde{x}$ be

$$\tilde{x} \in \mathbb{R}^9 := x - \bar{x}. \tag{5.19}$$

We start by defining the error set E as

$$E \triangleq \left\{ x \in \mathbb{R}^9 : X \oplus \{-\bar{x}\} \right\}, \tag{5.20}$$

where X is defined as

$$X \triangleq \left\{ x \in \mathbb{R}^9 : \theta \in \left(-\frac{\pi}{9}, \frac{\pi}{9}\right), \phi \in \left(-\frac{\pi}{9}, \frac{\pi}{9}\right), \psi \in \left(-\frac{\pi}{9}, \frac{\pi}{9}\right) \right\}, \tag{5.21}$$

ensuring an operation close to the linearized equilibrium.

The error dynamics are represented by

$$\dot{\tilde{x}}(\cdot) := \dot{x}(\cdot) - \underbrace{\dot{\bar{x}}(\cdot)}_{=0} \quad \Rightarrow \quad \dot{\tilde{x}}(\cdot) = \dot{x}(\cdot). \tag{5.22}$$

Remark 1 *We consider the velocity to be tracked by the MPC as constant through the whole receding horizon, as we sample the camera information only at the initial sampling time $\tilde{x}(k)$, which renders $\dot{\bar{x}}(n|k) = 0, \forall n = 0, ..., N$.*

Accordingly, we write the cost functions for the running cost $V_l(\cdot)$ and final cost $V_f(\cdot)$ to be used in the MPC formulation as

$$V_l(\tilde{x}(n|k), u(n|k)) = \tilde{x}(n|k)^T Q_x \tilde{x}(n|k) + u(n|k)^T Q_u u(n|k), \tag{5.23}$$

$$V_f(\tilde{x}(N|k)) = \tilde{x}(N|k)^T Q_f \tilde{x}(N|k). \tag{5.24}$$

where Q_x and Q_u are positive definite diagonal matrices that weight the state error and the control effort, respectively. These weights can be appropriately tuned to achieve a desired performance. Finally, the matrix Q_f is the final weight matrix for the last step of the receding horizon optimization, obtained through the Riccatti equation:

$$Q_f = Q_x + A^T Q_f A - (B^T Q_f A)^T (Q_u + B^T Q_f B)^{-1} (B^T Q_f A). \tag{5.25}$$

The final cost is defined through the Riccatti equation as it provides the best estimate of the cost-to-go for a receding horizon controller [58], similarly to the infinite horizon cost for Linear Quadratic Regulators [77].

The cost function, referred to as $J(\cdot)$, is defined as

$$J(\tilde{x}(0|k), u(0|k)) = \sum_{k=0}^{N-1} V_l(\tilde{x}(n|k), u(n|k)) + V_f(\tilde{x}(N|k)). \tag{5.26}$$

The optimal control input is then given by the following optimization problem

$$\min_{\mathbf{u}_k^*} \quad J(\tilde{x}(0|k), u(0|k)) \tag{5.27a}$$

$$\text{subject to} \quad \tilde{x}(n+1|k) = A\tilde{x}(n|k) + Bu(n|k), \tag{5.27b}$$

$$\tilde{x}(n|k) \in E, \tag{5.27c}$$

$$\tilde{x}(N|k) \in E_f, \tag{5.27d}$$

$$u(n|k) \in U \tag{5.27e}$$

$$\tilde{x}(0|k) = \tilde{x}(k) \tag{5.27f}$$

where E and U are given in (5.20), and (5.7), respectively.

Lastly, E_f is a terminal invariant set for $\tilde{x}(k)$. As terminal sets, one can choose the zero-set

$$E_f \triangleq \{\tilde{x} \in \mathbb{R}^9 : \tilde{x} = 0^9\} \subset E$$

or the Linear Quadratic Regulator (LQR) terminal set. To avoid the need for long horizon planning, which yields a high computational cost and has associated feasibility issues, we use the terminal set associated with a LQR controller.

We now state the theorem that yields the stability results for the proposed MPC controller.

Theorem 2 *Under the control law in (5.27), then $\tilde{x}(t)$ satisfies $\lim_{t\to\infty} \tilde{x}(t) = 0$ from all initial values of $\tilde{x}$ for which (5.27) admits a feasible solution.*

Proof: The proof of this theorem follows the corresponding analysis in [58] and is divided in three parts: at first, we prove that the terminal set E_f is a control invariant set; at the second, we demonstrate the boundedness of $J_N^0(\tilde{x})$; and finally we guarantee that $J_N^0(\tilde{x})$ is monotonically decreasing.

Following [58], let $J_N^0(\tilde{x}) = J(\tilde{x}(0|t), u_N(\cdot|t))$, where $\tilde{x}(0|t) \in E$ is any MPC starting state and $u_N(\cdot|t) = \{u_N(0|t), u_N(1|t), \ldots, u_N(N-1|t)\}$ be a minimizing control sequence.

1) Control Invariant Sets: The controllability matrix for system (5.9) is of the form

$$C_m = \begin{bmatrix} A^n B & A^{n-1} B & \dots & AB \end{bmatrix} \in \mathbb{R}^{9 \times 36}, \tag{5.28}$$

where A and B are defined in (5.15) and (5.16).

In our case, the rank of C_m is 9, and the condition number of C_m is approximately 388, for our system properties defined in Section 5.4. Therefore, we conclude that the system (5.9) is controllable, and therefore stabilizable. As so, there exists a terminal feedback controller $u(N|k) = -\Gamma \tilde{x}(N|k)$ that renders E control invariant.

Remark 2 *The condition number of C_m is dependent on the system inertial parameters that affect the control matrix B. Therefore, it is worth noting that these results are valid for the particular system in hand.*

2) Recursive Feasibility: To prove recursive feasibility, we recall that E is a control invariant set, that is, $\forall \tilde{x} \in E$, there exists at least one $u \in U$ such that $f(\tilde{x}, u) \in E$. Note also that $\tilde{x}(t) = \tilde{x}(0|k)$.

The control law in (5.27) gives an optimal control signal for the system in (5.19), yielding $\tilde{x}(n + 1|k) = f(\tilde{x}(n|k), u(n|k))$. The control inputs through the receding horizon, then, satisfy the constraints set by E and U if and only if the problem is initially feasible.

At step $n = N$, $f(\tilde{x}(N|k), u_f(N|k)) = \tilde{x}(N + 1|k) \in E_f$, as there exists at least one control input $u_f(N|k)$, that renders the set E_f control invariant, *i.e.*, $\tilde{x}(N+1|k) \in E_f$ for at least one $u_f(N|k)$ control input. Such control input can be a linear feedback controller $u_f(N|k) = -\Gamma \tilde{x}(N|k)$, and Γ is an appropriately selected gain matrix (for instance, LQR feedback). As so, if the optimization problem in (5.27) is feasible at t, then it will be recursively feasible for $t + \Delta t$. □

3) Boundedness of $J_N^0(\tilde{x})$: The bounds of $J_N^0(\tilde{x})$ are derived from the choice of the running cost function, given by $\tilde{x}(n|k)^T Q_x \tilde{x}(n|k) + u(n|k)^T Q_u u(n|k)$, with Q_x and Q_u positive definite. Therefore, it holds that

$$\alpha_1(||\tilde{x}||, ||u||) \leq J_N^0(\tilde{x}) \leq \alpha_2(||\tilde{x}||, ||u||) \tag{5.29}$$

where $\alpha_1(||\tilde{x}||, ||u||) = \underline{\lambda}(Q_x)||\tilde{x}|| + \underline{\lambda}(Q_u)||u||$ and $\alpha_2(||\tilde{x}||, ||u||) = \bar{\lambda}(Q_x)||\tilde{x}|| + \bar{\lambda}(Q_u)||u||$, and where $\underline{\lambda}(Q_i)$ and $\bar{\lambda}(Q_i)$ are, respectively, the minimum and maximum eigenvalues of matrix Q_i. Noting that Q_x and Q_u are diagonal positive-definite matrices, then the eigenvalues are displayed on the diagonal of these matrices and correspond to the appropriate weights chosen for the cost function. For the experiments we chose $\lambda_{max}(Q_x) = 100, \lambda_{min}(Q_x) = 1$ and $\lambda_{max}(Q_u) = 2, \lambda_{min}(Q_x) = 1$, leading to

$$||\tilde{x}|| + ||u|| \leq J_N^0(\tilde{x}) \leq 100||\tilde{x}|| + 2||u|| \tag{5.30}$$

4) Descent and Monotonicity Properties of the Cost Function: The remainder of this proof follows the same steps as in [58], Proposition 2.12, proof for Theorem 2.24.

Finally, it can be concluded that the system in (5.19), under the control law in (5.27), converges to the origin, that is, the error $\tilde{x}(t)$ asymptotically converges to zero, and therefore the system achieves the desired tracking of the reference velocity from the IBVS module.

5.4 Experiments

We run two different types of experiments. We start with realistic simulation results using Gazebo and the RotorS simulator [78] (experiments presented in Sec. 5.4). No disturbances are considered and a good approximation of the system model is used. Then, in Sec. 5.4, we tested the proposed framework in a real experimental scenario using a UAV with very little knowledge of the system's model.

For both simulations and real experiments, we perform the same excitation, with desired features setpoints, to the system: sequential translations of $0.5[m]$ on x, y and z axis, achieved by changing the desired image features f_i^* accordingly. We have generated C/C++ code using CVXGEN [79], with an horizon of $N = 10$, for the Linear MPC implementation of (5.27), which was wrapped around the ROS framework [80]. Implementation-wise, both simulations and real experiments share the same ROS backend, apart from the vehicle interface. In simulation, we rely on RotorS, while for the real experiments we use a PX4-based UAV [76] and the MAVROS package[1] to interface with the flight controller.

Simulation Results

Figure 5.3 shows the results obtained in simulation. Figure 5.3(a) shows the feature error and 5.3(b) the IBVS velocity convergence. On the 3D setup, the UAV starts at an approximate distance to the target of $1.3[m]$, composed by 4 features. The simulated vehicle has a mass of $1.5[kg]$ and inertia of $M_{[x]} = 0.034, M_{[y]} = 0.046$ and $M_{[z]} = 0.098$.

As expected, with a good knowledge of the system parameters, our control method achieves a practically zero steady-state error, converging asymptotically to the desired reference with a small overshoot on the z axis. This result is important in assessing the proposed scheme, as it means that the connection of the IBVS module and the MPC works properly, even under the assumptions made. The MPC solver took a minimum of $0.72[ms]$, a maximum of $5.8[ms]$, and an average of $1.1[ms]$ to obtain a solution on a $3.6[GHz]$ Intel Core-i7 CPU.

Experimental Results

On what concerns the experimental test-bed, our UAV, built around a Hover 1 frame, a Nvidia Jetson TX2, and a PX4 flight controller has an approximate mass of $1.73[kg]$ and a rough inertia approximation of $M_{[x]} = 0.04, M_{[y]} = 0.04$, and

---

[1]Available at (on 12th September, 2019): `https://github.com/mavlink/mavros`

$M_{[z]} = 0.1$. To obtain the depth Z of each feature, we use a ZED Stereo camera. The system successfully converges to the desired setpoints, although a small steady-state error is present. The obtained 3D trajectory, as shown in Figure 5.5, provides an insight on the system's performance while navigating through the different setpoints. On the vehicle, running a Nvidia Jetson TX2, the solver took a minimum of $7.9[ms]$, a maximum of $10.8[ms]$ and an average of $8.4[ms]$ to obtain a sequence of control inputs, making it possible to run the controller at $100[Hz]$.

We observe a small decrease of the tracking performance on the real system. This results from (i) an imperfect knowledge of our system inertial parameters, and (ii) an inaccurate force-to-thrust mapping, dependent on motor properties and propellers performance. In typical PID implementations, the integrator gain compensates for these system errors, converging the steady-state error to zero. Nonetheless, a PID approach is subject to high gain tuning and input saturation that can render the system unstable, and it cannot deal with state constraints. At this stage, we did not add an integration action to the MPC controller. Nonetheless, the system coped with the state constraints that avoid states far from the linearization point, and kept the desired small attitude angles during the entire test sets. Moreover, the computational benefits with respect to previous approaches of using MPC for quadrotors, e.g. [75], make this approach easily implementable on-board UAVs of different sizes and computational capabilities, ensuring an operation near the linearization point of the system.

Furthermore, the use of the MPC framework was crucial to limit the velocity of the vehicle and cope with motion blur that the used camera is subject to.

5.5 Discussion

We present a new formulation to tackle the under-actuation issues of the image-based visual servoing of a quadrotor. The guarantees on the feasibility and stability of the MPC are presented. Using both synthetic and real data, we show our method works properly and is capable of running at a frequency of 100Hz on-board the vehicle. As future work, we plan to extend the formulation to a Linear Time-Varying MPC to handle time-varying dynamics - mainly, due to the changing velocity setpoint from the IBVS module, and incorporate the visual servoing dynamics in the MPC framework.

In the next chapter, a MPC formulation taking into account the feature dynamics in the model is proposed, so that the IBVS and MPC module can be combined into one.

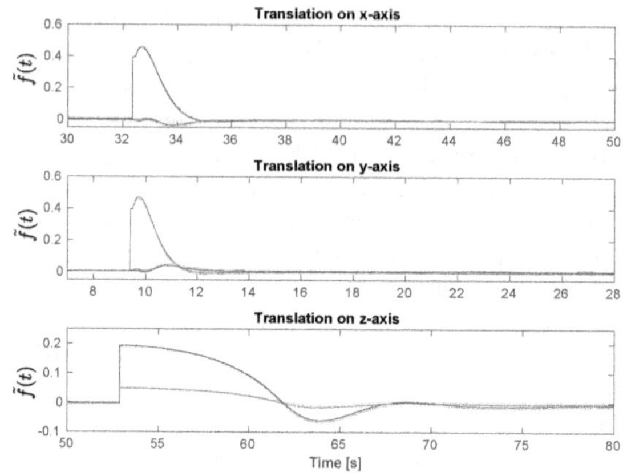

(a) Feature Error. Each curve represents a feature component error (x and y in the image plane).

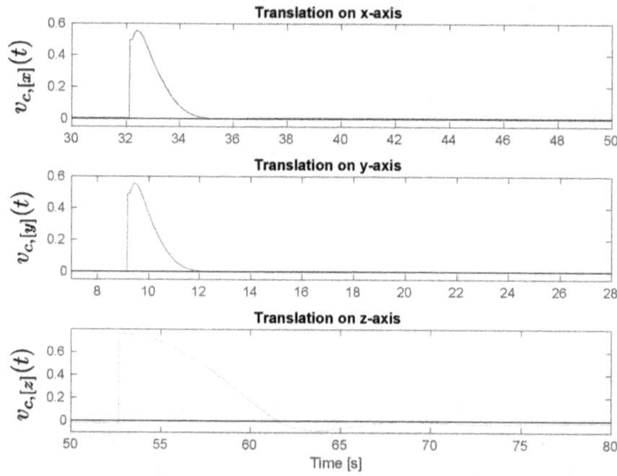

(b) IBVS Velocity Module. The curves in blue, red and yellow represent the desired velocity along the x, y and z axis.

Figure 5.3: Simulation results: We use a realistic physics engine simulator, namely Gazebo, and RotorS [78]. We observe that the feature errors converge to zero and, accordingly, so does the desired velocity for the camera/UAV. Note that convergence on the x and y axis is faster (and similar among each other) than the convergence on the z axis, due to the system dynamics. On Fig. (a), each curve corresponds to a feature error. On Fig. (b), the zero error is shown in black, while the velocity for x, y and z is shown in blue, red and yellow.

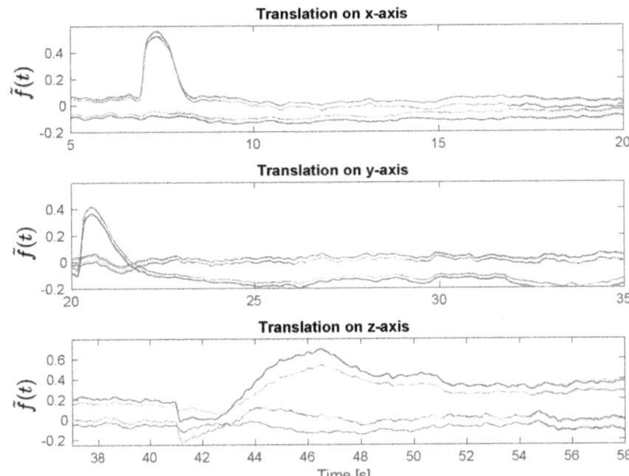

(a) Feature Error for x, y and z translations, top to bottom. Each curve represents a feature component error (x and y in the image plane).

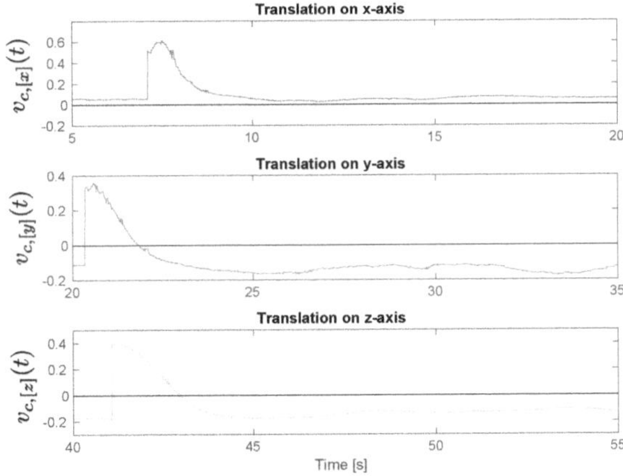

(b) IBVS Velocity Module. The curves in blue, red and yellow represent the desired velocity along the x, y and z axis, respectively.

Figure 5.4: Performance of the real system: We observe that, although the system converges to the correct desired feature position in the image plane, there is a small steady-state error. This error is due to the imperfect knowledge of the system's inertial properties. This is observed on the error around the z-axis, as a force-to-thrust mapping error causes a small decrease in the system performance. Nonetheless, we observe that the vehicle navigates stably through the desired feature sets.

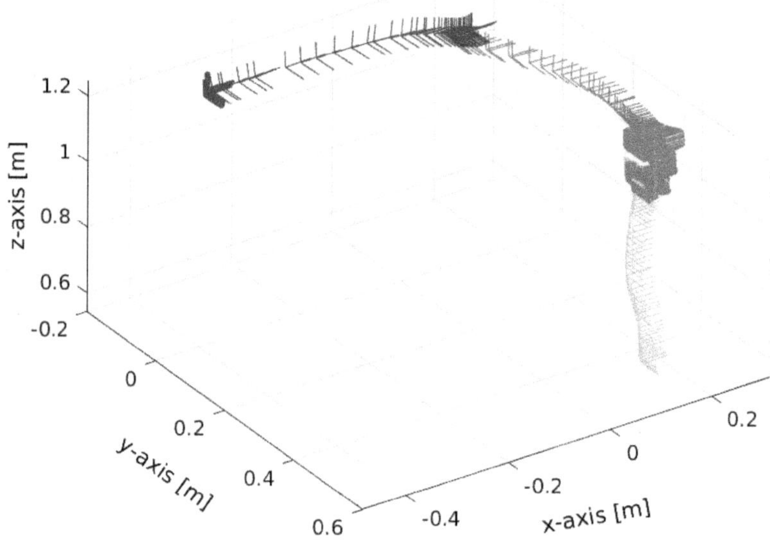

Figure 5.5: 3D Trajectory of the UAV. Blue, red and yellow represent the translations on x, y and z respectively.

Chapter 6

Vision-based Predictive Control for Coordination of Multi-Agent Formations

Coordinating a group of agents in underground or indoor environments is a hard task. Precise localization systems are often expensive and proprioception is often not accurate enough for precise formation control, which tasks such as distributed mapping, load transportation and distributed aperture imaging systems often require. In this chapter, we propose a vision-based formation control approach, that relies on image features collected from onboard cameras to coordinate multiple agents operating in the same environment, and is validated both in simulation and experiments. We show that combining visual feedback in the control loop can provide reliable performance for formation keeping and coordination tasks.

6.1   Literature review

This work focuses on a multi-agent formation control problem (Fig. 6.1) for M agents, one leader and $M - 1$ followers. The objective is to derive control laws for the followers to converge and maintain predefined relative poses between each other and the leader. While this is a well-studied problem, there are still several challenges related to the absence/limited availability of accurate relative pose sensing, such as in underground or indoor environments. To cope with this, we explore the use of image features from on-board cameras in the control module. We propose two methods for multi-agent relative pose coordination that guide the followers' agents to the desired relative position. The methods here proposed use Model Predictive Control (MPC) [81–83] and Visual Servoing [57,84] which, by matching features in the images of each agent, and using one range sensing measurement from the leader to one follower, provide control inputs to achieve a desired formation geometry.

Image-based Visual Servoing (IBVS) is a well-known technique for driving an

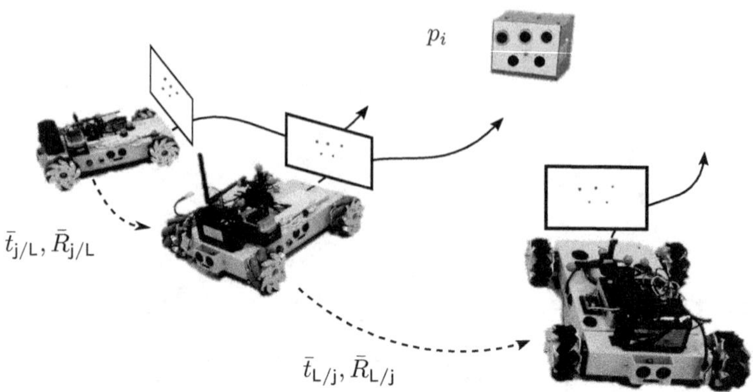

Figure 6.1: Formation with 3 agents. The formation is defined as relative positions and attitudes with respect to the camera frames, which can be translated into the agent's frames.

agent to a desire position [57,84]. Instead of controlling the position of the robot directly, IBVS models the agent's velocities as a function of the errors between the current and goal image features from on-board cameras. The first works exploring epipolar geometry in the IBVS are presented in [85,86]. The authors assume the environment is unknown, making it impossible to set goal image features. They define the errors as the distance between the image features and the epipolar lines, obtained from the current and desired relative pose (see [14]), and their method drives the robot to a desired position, up to a scale factor. We address instead here the problem of minimizing both rotation and translation at the same time, and include control and state constraints to ensure the features remain in the image.

On what concerns omnidirectional cameras, multiple works have addressed the representation of these imaging systems under unified frameworks. In [87–89], generalized camera systems are modeled to represent imaging sensors that can range from projective cameras, through parabolic, catadioptric and radial distortion (*fisheye*). In [15], generalized camera models are used to formulate the epipolar constraint for general camera systems in a linear fashion, by lifting the imaging plane to 5 dimensions via Veronese maps. On what regards visual-servoing for general camera systems, in [19] the authors introduce an interaction matrix for generalized cameras.

Regarding relative pose coordination using image features, [90] derives a method that provides control inputs from the epipoles computed from neighboring robots. The method will reach consensus in their orientations, without the need of directly observing each other. In [91], the authors use IMU and computer vision to obtain a rectified image [14]. The method is appropriate for aerial vehicles with down-pointing cameras. A distributed consensus scheme to deal with the translation scale is proposed. Other vision-based approaches for relative pose coordination are available: [92, 93] present methods for motion coordination and control strategy for leader-follower formations of non-holonomic vehicles, under visibility and communication constraints, as well as saturation of control inputs. [94] uses a distributed consensus of $M \geq 3$ agents for aerial-robotic teams. Using a PID-based control, each robot uses its view of a target and the relative distance from its two closest neighbors. [95] addresses the formation control of aerial vehicles with downward-facing cameras. The solution computes the control commands from the projection of a subset of ground vehicles. [96] presents a vision-based method for incremental depth and relative pose estimation of ground vehicles.

We propose here a novel Image-Based Formation Control (IBFC) framework which employs visual information to drive the robotic agents to maintain a desired formation. The goal is to generate control inputs using corresponding image features from the robots' on-board cameras. In contrast to [90, 91], we i) locally obtained control inputs in the image-frame, without the need for global localization methods or heading references, such as those provided by a magnetometer; ii) allow for the 6 Degrees-of-Freedom (DoF) formation coordination with multiple camera types, by modeling each camera as a general projection system; and iii) use an MPC framework able to generate optimal control inputs in the image space. In contrast to [94] that uses range sensing between the current and at least two neighboring agents, we only require one range distance, whereas [95] addresses an entirely different problem, as it uses ground vehicles as landmarks, limiting severely the number of features that can be used. We stress that, while some previous methods explore epipolar lines for IBVS, to the best of our knowledge, this is the first approach that exploits these constraints for multi-agent formation control. Moreover, the use of intersections from multiple epipolar geometric constraints for $M > 2$ agents formation control is also new in the literature, allowing us to avoid the use of relative position measurements for large formations, extending considerably the application areas of our method with respect to the state of the art.

6.2 Dynamics

The formation geometry is defined by the desired relative poses among the agents, that is, $\bar{R}_{j/l}$ and $\bar{t}_{l/j}$, $j, l \in M$, with $j \neq l$. Fig. 6.1 depicts a formation with desired relative positions and orientations.

We consider two types of followers - one, here denominated F_1, that can measure a distance to the leader L, with state

$$\xi^{\mathsf{F}_1} = \begin{bmatrix} \|t_{\mathsf{F}_1/\mathsf{L}}\| & f_1^{\mathsf{F}_1} & \cdots & f_5^{\mathsf{F}_1} \end{bmatrix}^T \in \mathbb{R}^{18}, \tag{6.1}$$

and kinematics given by

$$\dot{\xi}^{\mathsf{F}_1} = \begin{bmatrix} \frac{t_{\mathsf{F}_1/\mathsf{L}}^T}{\|t_{\mathsf{F}_1/\mathsf{L}}\|} \begin{bmatrix} -I_{3\times3} & -t_{\mathsf{F}/\mathsf{L}_\times} \end{bmatrix} \\ L_1(f_1^{\mathsf{F}_1}) \\ \vdots \\ L_5(f_5^{\mathsf{F}_1}) \end{bmatrix} u^{\mathsf{F}_1}. \tag{6.2}$$

The remaining followers and leader, that only measure image-features, have their states given by

$$\xi^r = \begin{bmatrix} f^r & \cdots & f_5^r \end{bmatrix}^T \in \mathbb{R}^{15}, j = 2, ..., \mathrm{M}, \tag{6.3}$$

with kinematics

$$\xi^r = \begin{bmatrix} L_1(f_1^r) \\ \vdots \\ L_5(f_5^r) \end{bmatrix} u^{\mathsf{F}}. \tag{6.4}$$

with $r = \{\mathsf{L}, \mathsf{F}_2, \ldots, \mathsf{F}_M\}$. In the above, L_i are the interaction matrices defined in (2.23). Note that these matrices also depend on the feature depth z_i, which is updated at every sampling time, as suggested in [97]. Each state ξ^r is constrainted by a polytope Ξ_o where we wish the state to evolve, that is

$$\xi^r \in \Xi_r. \tag{6.5}$$

In the same way, the control input is constrained as $u^r \in \mathbb{R}^6 \subset U$, where U is defined as

$$U \triangleq \{u^j \in \mathbb{R}^6 : u_{[q]}^{\min} \le u_{[q]}^j \le u_{[q]}^{\max}\}, q = 1, \ldots, 6. \tag{6.6}$$

6.3 Problem Statement

In this work, we aim at exploit the epipolar geometry to define the formation control problem, enabling us to obtain a desired formation configuration based solely on image features f_i^j and f_i^l, and a relative distance measurement $\|t_{\mathsf{j}/\mathsf{l}}\|$ between two agents in M. Formally, we define the problem as follows:

Problem 2 *Given a set of essential matrices $\{\tilde{E}_j^l\}$, encoding the desired relative pose between two agents $j, l \in \mathrm{M}, j \ne l$, up to a scale factor (2.20), a relative distance measurement $\|t_{\mathsf{j}/\mathsf{l}}\|$ between two agents[1], two sets of features $F^j = \{f_1^j, \ldots, f_i^j\}$ and $F^l = \{f_1^l, \ldots, f_i^l\}$, observed by agents $j, l \in \mathrm{M}$ respectively, with at least 5 feature correspondences, design $u^l, r = \{\mathsf{F}_1, \mathsf{F}_2, \ldots, \mathsf{F}_M\}$ such that the agent l camera system in (6.2) or (6.4), under constraints (6.5) and (6.6), is driven to a desired relative attitude $\bar{R}_{\mathsf{j}/\mathsf{l}}$ and relative position $\bar{t}_{\mathsf{j}/\mathsf{l}}$.*

---

[1]This relative distance can be measured through ultrasonic or ultra-wide band sensors.

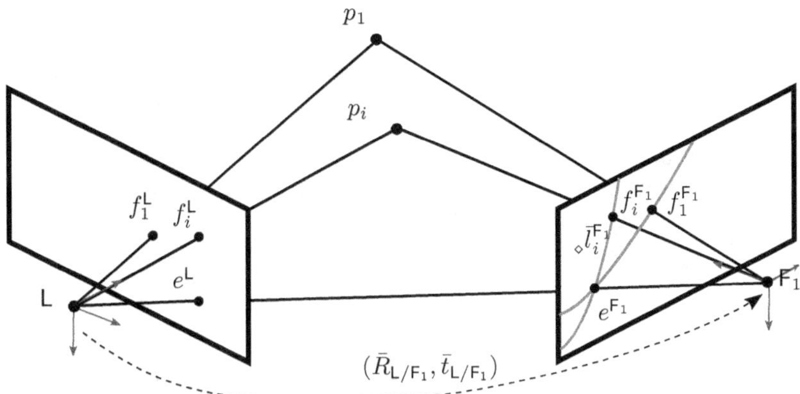

Figure 6.2: Epipolar geometry for the formation control scenario. The formation is defined by a desired relative pose parameterized by $\bar{R}_{\mathsf{L}/\mathsf{F}_1}$ and $\bar{t}_{\mathsf{L}/\mathsf{F}_1}$ between the leader L and the follower F_1. The epipolar constraint states that, in the F_1 image plane, $f_i^{\mathsf{F}_1}$ must lie on the respective epipolar curves ${}_\diamond l_i^{\mathsf{F}_1}$ (red lines).

In the next two chapters we present the $\mathrm{M} = 2$ and $\mathrm{M} > 2$ agent relative pose coordination control schemes.

6.4 Two Agent Coordination

We first consider the problem of relative pose control between L and one follower (F_1), assuming that F_1 has a relative range measurement with respect to L. This corresponds to the $\mathrm{M} = 2$ scenario. We abbreviate F_1 to F in the sequel. Consider (2.20) and the scenario in Fig. 6.2, that can be extended to an arbitrary number of points p_i^{W} - with corresponding features f_i^{L} and f_i^{F}, on L and F, respectively. In the same way, each desired feature position $\bar{f}_i^{\mathsf{L}}$ generates an epipolar curve ${}_\diamond l_i^{\mathsf{F}}$ in the camera frame of the follower through

$$
{}_\diamond \bar{l}_i^{\mathsf{F}} = \tilde{\bar{E}}_{\mathsf{F}}^{\mathsf{L}} \, {}_\diamond \bar{f}_i^{\mathsf{L}}. \tag{6.7}
$$

To achieve the desired pose – up to a scale factor in translation – five points are needed [98], $p_i^{\mathsf{W}}, i = 1, \ldots, 5$, each of which generating five epipolar curves ${}_\diamond \bar{l}_i^{\mathsf{F}}, i = 1, \ldots, 5$. As it is assumed that the camera is aligned with the body frame of the agents, the control input to the system is $u^{\mathsf{F}} \in \mathbb{R}^6$, corresponding to the follower's camera velocity, as in (2.25).

Model Predictive Image-Based Visual Servoing

To align the observed points f_i^{F} with the curves ${}_\circ \bar{l}_i^{\mathsf{F}}, i = 1, \ldots, 5$, and to keep a desired distance between the leader and follower, we employ a Nonlinear Model Predictive Controller with image feedback - hereafter referred to as Model Predictive Image-Based Formation Controller (MP-IBFC). MP-IBFC is a Finite-Horizon Optimal Controller (FHOC) [58] which minimizes a cost function $J(\cdot)$, that depends on the state and control, along a receding horizon of length N, while taking into account the system model. The MP-IBFC is defined with respect to an appropriate error state. The goal is 2-fold: i) minimize the error between the current and desired range from the leader to the follower; and ii) minimize the distance of the features to the corresponding epipolar curves.

On the lifted Veronese space, a feature ${}_\circ f_i^{\mathsf{F}}$ belongs to the curve ${}_\circ \bar{l}_i^{\mathsf{F}^T}$ if their inner product is zero, that is, ${}_\circ \bar{l}_i^{\mathsf{F}^T} {}_\circ f_i^{\mathsf{F}} = 0$. However, if such a point does not belong to the latter, then there exists an algebraic error corresponding to ${}_\circ \bar{l}_i^{\mathsf{F}^T} {}_\circ f_i^{\mathsf{F}} \neq 0$. This algebraic error ultimately belongs to a 2D plane (the image plane), and therefore it may be written as $f_i^{\mathsf{F}^T} M_i f_i^{\mathsf{F}} = 0$, where $M_i = \frac{1}{2} \begin{bmatrix} 2a & b & d \\ b & 2c & e \\ d & e & 2 \end{bmatrix}$ and considering that ${}_\circ \bar{l}_1^{\mathsf{F}} = \begin{bmatrix} a & b & c & d & e & 1 \end{bmatrix}$ represents the lifted curve, for some constants a through e. Minimizing all the algebraic errors $f_i^{\mathsf{F}^T} M_i f_i^{\mathsf{F}} = 0$ leads to $R_{\mathsf{L/F}} = \bar{R}_{\mathsf{L/F}}$ and $t_{\mathsf{L/F}} \sim \bar{t}_{\mathsf{L/F}}$, that is, equality up to a scale factor. To obtain $t_{\mathsf{L/F}} = \bar{t}_{\mathsf{L/F}}$, a distance measurement $\|t_{\mathsf{L/F}}\|$ between the two agents is needed, such that the scale factor is accounted for and the two agents achieve the desired formation.

Accordingly, the errors are defined, in Cartesian coordinates, as

$$\epsilon = \begin{bmatrix} \|t_{\mathsf{F/L}} - \bar{t}_{\mathsf{F/L}}\| \\ \frac{1}{2} f_1^{\mathsf{F}^T} M_1 f_1^{\mathsf{F}} \\ \vdots \\ \frac{1}{2} f_5^{\mathsf{F}^T} M_5 f_5^{\mathsf{F}} \end{bmatrix} \in \mathbb{R}^6 \subset \mathcal{E} \tag{6.8}$$

where $\|t_{\mathsf{F/L}} - \bar{t}_{\mathsf{F/L}}\|$ represents the relative distance error, and $f_i^{\mathsf{F}^T} M_i f_i^{\mathsf{F}}$ represents the algebraic curve-distance errors, for $i = 1, \ldots 5$. The set $\mathcal{E}$ that contains the state-errors is defined as

$$\mathcal{E} \triangleq \{\epsilon \in \mathbb{R}^6 : \epsilon_{[r]}^{\min} \leq \epsilon_{[r]} \leq \epsilon_{[r]}^{\max}\} \tag{6.9}$$

where the minimum and maximum allowed errors, respectively $vec\epsilon_{[r]}^{\min}$ and $vec\epsilon_{[r]}^{\min}$, are obtained from the maximum allowed distance error in the agents workspace and the maximum allowed error in the camera image frame.

From (6.8), we derive the error dynamics as

$$
\dot{\epsilon} = \tilde{L}u^{\mathsf{F}} =
\begin{bmatrix}
\frac{t_{\mathsf{F/L}}^{T}}{\|t_{\mathsf{F/L}}\|}\begin{bmatrix} -I_{3\times 3} & -t_{\mathsf{F/L}\times} \end{bmatrix} \\
\left[f_1^{\mathsf{F}^{T}} M_1 \right]_{[1:2]} L_1(f_1^{\mathsf{F}}) \\
\vdots \\
\left[f_5^{\mathsf{F}^{T}} M_5 \right]_{[1:2]} L_5(f_5^{\mathsf{F}})
\end{bmatrix} u^{\mathsf{F}},
\tag{6.10}
$$

where, $\left[f_i^{\mathsf{F}^{T}} M_i \right]_{[1:2]}$ is the first to second columns of the vector $f_i^{\mathsf{F}^{T}} M_i$ - as the feature dynamics are given by $\dot{f}_i^{\mathsf{F}} = \begin{bmatrix} \dot{u}_i & \dot{v}_i & 0 \end{bmatrix}$. Note that $t_{\mathsf{F/L}}$ is obtained from the range measurement and the desired translation vector encoded in the desired essential matrix $\tilde{E}_{\mathsf{F}}^{\mathsf{L}}$, which is a good estimate of the actual relative position between the agents for small relative attitude errors (that are minimized in the control law with $f_i^{\mathsf{F}^{T}} M_i f_i^{\mathsf{F}}$).

MP-IBFC is implemented in discrete-time with a sampling time Δt. The sampled error dynamics are given by

$$
\epsilon(k+1) = g_\epsilon(\xi^{\mathsf{F}}(k), u^{\mathsf{F}}(k)) = \epsilon(k) + \Delta t \tilde{L}u^{\mathsf{F}}(k), \forall k \in \mathbb{N}_{\geq 0},
\tag{6.11}
$$

and $\xi^{\mathsf{F}}(k)$ is given as in (6.2).

Finally, we define the cost function to minimize, $J(\epsilon(k), u^{\mathsf{F}}(k))$, as

$$
J(\epsilon(k), u^{\mathsf{F}}(k)) = \sum_{n=0}^{N-1} \epsilon(n|k)^{T} Q_e \epsilon(n|k) + u^{\mathsf{F}^{T}}(n|k) Q_u u^{\mathsf{F}}(n|k)
$$
$$
+ \epsilon(N|k)^{T} Q_N \epsilon(N|k)
\tag{6.12}
$$

where Q_e and Q_u are positive-definite weighing matrices, and Q_N is a positive definite weight matrix associated with a terminal invariant set $\mathcal{E}_f = \{\epsilon \in \mathbb{R}^8 : \epsilon^{T} Q_N \epsilon \leq \delta\}, \delta \in \mathbb{R}_+$, where for $\epsilon(k) \in \mathcal{E}_f$, we assume that there exists a controller $u^{\mathsf{F}} \in U$ such that $\epsilon(k+1) \in \mathcal{E}_f$. Let us denote

$$
l(\epsilon, u^{\mathsf{F}}) = \epsilon(n|k)^{T} Q_e \epsilon(n|k)
$$
$$
+ u^{\mathsf{F}^{T}}(n|k) Q_u u^{\mathsf{F}}(n|k) \text{ and}
\tag{6.13}
$$
$$
V(\epsilon) = \epsilon(N|k)^{T} Q_N \epsilon(k+N|k).
\tag{6.14}
$$

The complete MP-IBFC formulation is then given by

$$\min_{u^{\mathsf{F}}} \; J(\epsilon(k), u^{\mathsf{F}}(k)) \tag{6.15a}$$

$$\text{subject to: } \epsilon(n+1|k) = g_\epsilon(\xi(n|k), u^{\mathsf{F}}(n|k)), \tag{6.15b}$$

$$\epsilon(n|k) \in \mathcal{E} \tag{6.15c}$$

$$u^{\mathsf{F}}(n|k) \in U \tag{6.15d}$$

$$n = 0, \ldots, N-1 \tag{6.15e}$$

$$\epsilon(0|0) = \epsilon(0). \tag{6.15f}$$

Note that the algebraic curve-distance errors, $\frac{1}{2} f_i^{\mathsf{F}^T} M_i f_i^{\mathsf{F}}, i = 1, \ldots 5$, are non-bijective - that is, it is possible to obtain the error ϵ knowing features f_i^{F}, but it is not possible to obtain each feature f_i^{F} given the error ϵ. Therefore, we derive the following result, based on the derivations in [99].

Theorem 3 *Under the control law in (6.15) and the assumptions of Problem 2, (6.10) is asymptotically stabilized to the origin.*

Proof: From [99], (6.15) asymptotically stabilizables (6.10) to the origin if two assumptions are met: i) the terminal cost (6.14) is a Control Lyapunov Function (CLF) of (6.10) with an associated terminal set $\Omega = \{\epsilon \in \mathbb{R}^6 : V(\epsilon) < \delta\}, \delta \in \mathbb{R}_+$ such that

$$\alpha_1(\|\epsilon\|) \leq V(\epsilon) \leq \alpha_2(\|\epsilon\|) \tag{6.16}$$

$$\min_{u \in U} \{V(g_\epsilon(\epsilon, u)) - V(\epsilon) + l(\epsilon, u)\} \leq 0 \tag{6.17}$$

In [100], two approaches are suggested to derive such CLFs and respective invariant sets: i) [12], for continuous-time MPC, and ii) [101] fro discrete-time MPC formulations. However, these approaches derive such sets from linear feedback controllers, which cannot be applied for the dynamics in (6.10) due to the non-bijectivity of the error. Therefore, we propose an approach to derive such sets using nonlinear feedback controllers.

Consider (6.10). Note that $\tilde{L}$ is full rank for a proper selection of image features which are not co-linear [57], and therefore we can calculate its inverse, defined as $\tilde{L}^\dagger$. Considering (6.11) we propose a local terminal controller in Ω given by

$$u^{\mathsf{F}}(k) = -\frac{\tilde{L}^\dagger}{h} S\epsilon(k), \tag{6.18}$$

where S is a gain matrix designed on the forthcoming. Under the control action (6.18), the dynamics in (6.11) can be written as

$$\epsilon(k+1) = \underbrace{(I - S)}_{\Phi} \epsilon(k) \tag{6.19}$$

where S is designed such that the eigenvalues of Φ are within the unit circle. Then, we may write $V(\epsilon(k+1)) = \epsilon(k)^T \Phi^T Q_N \Phi \epsilon(k)$ and we wish to design Q_N such that

$$V(\epsilon(k+1)) - V(\epsilon(k)) = -\epsilon(k)^T Q^* \epsilon(k) \tag{6.20}$$

where $Q^* = Q_e + \frac{1}{h^2} S^T Q_u S + \Delta Q$, for any positive definite ΔQ. To obtain such Q_N, we calculate the solution to the discrete Lyapunov equation

$$\Phi^T Q_N \Phi - Q_N = Q^*. \tag{6.21}$$

At this point, we will attempt at obtaining the constant δ that defines the Ω set. First, note that we may upper-bound $\|\tilde{L}\|$ with a constant $\bar{L}, \bar{L} > 0$ such that $\|u^{\mathsf{F}}\| \leq \bar{L} \cdot \|S\| \cdot \|\epsilon\|$ due to i) M_i being a constant matrix during the operation, $i = 1, \dots, 5$; ii) the image features being limited to the image-space in normalized coordinates, respecting (6.5); and iii) the range $t_{\mathsf{F}}^{\mathsf{L}}$ being upper-bounded due to the state constraints in (6.5). Then, note that $\lambda_{min}(Q_N)\|\epsilon\|^2 \leq \epsilon Q_N \epsilon < \delta$, where $\lambda_{min}(Q_N)$ is the smallest eigenvalue of Q_N. Then, for a $\delta < \lambda_{min}(Q_N)\left(\frac{\|\bar{u}\|}{\|S\|\bar{L}}\right)^2$, where $\bar{u} = [u_1^{max}, \dots, u_6^{max}]^T$, we obtain that $\lambda_{min}(Q_N)\|\epsilon\|^2 < \delta \Rightarrow \lambda_{min}(Q_N)\|\epsilon\|^2 < \lambda_{min}(Q_N)\left(\frac{\|\bar{u}\|}{\|S\|\bar{L}}\right)^2$. Then, [99, Assumption 1] is fulfilled.

Lastly, we require the running cost (6.13) to be upper and lower bounded by constants d and D, respectively. Noting that Q_e and $\mathbb{Q}_u$ are positive definite, we may write $d \leq l(\epsilon, u) \leq D$ where $d = \lambda_{\min}(Q_e)\|\epsilon\| + \lambda_{\min}(Q_u)\|u\|$ and $D = \lambda_{\max}(Q_e)\|\epsilon\| + \lambda_{\max}(Q_u)\|u\|, \forall \epsilon \in E$ and $u \in U$, fulfilling [99, Assumption 2].

By fulfilling [99, Assumption 1] and [99, Assumption 2], and from Lemma 1 and Theorems 1-3 in [99], the result is obtained, thus completing the proof. $\square$

6.5 Formation Control

Assuming that at least one follower is in formation with the leader (without loss of generality, let it be F_1) using the MP-IBFC controller proposed in 6.4, we propose a control scheme that is capable of driving an agent to the correct relative pose in the formation based solely on image features.

Consider the scenario in Fig. 6.3, where agents L and F_1 are at the desired relative pose, and follower F_j receives image features from both agents. In this setting, we show that the relative pose of the follower F_j with respect to L and F_1 is defined by the epipolar geometry shared by the three camera system. This property can be extended for any agent $\mathsf{F}_j, j = 2, \dots, \mathsf{M} - 2$.

Let $_{\diamond}f_i^{\mathsf{F}_j}, i = 1, \dots, 5$, be the set of features observed by the follower $\mathsf{F}_j, j = 2, \dots, \mathsf{M} - 2$, and epipolar curves sets $_{\diamond}\bar{L}_{\mathsf{L}}$ and $_{\diamond}\bar{L}_{\mathsf{F}_1}$ received from the leader L and follower F_1, respectively. The relative pose, up to a scale factor, that minimizes the error between $_{\diamond}f_i^{\mathsf{F}_j}$ and the curve sets $_{\diamond}\bar{L}_{\mathsf{L}}$ and $_{\diamond}\bar{L}_{\mathsf{F}_1}$ is the one where all features in $_{\diamond}f_i^{\mathsf{F}_j}$ lie on top of the respective curves in $_{\diamond}\bar{L}_{\mathsf{L}}$ or $_{\diamond}\bar{L}_{\mathsf{F}_1}$. However, with two sets of curves, which represent the desired relative pose to the leader L and follower F_1,

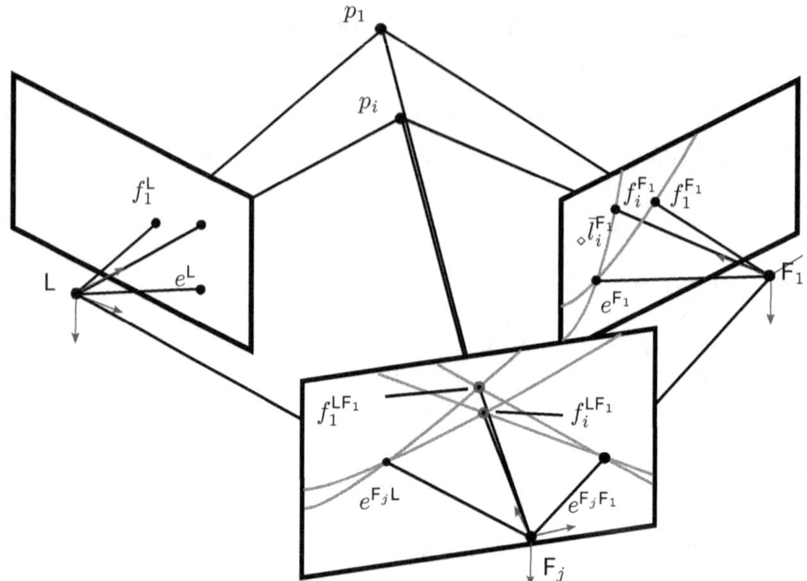

Figure 6.3: Depiction of the epipolar geometry for the formation control scenario with 3 or more agents. Considering two sets of desired relative poses parameterized by $\bar{R}_{j/m}$ and $\bar{t}_{j/m}$, and $\bar{R}_{l/m}$ and $\bar{t}_{l/m}$, $j, l, m \in M$, such that for the leader L, and two followers F_1 and F_j (j different than L and F_1), it is possible to define epipolar constraints that define the relative pose between the three agents from image measurements, without the need of external relative position systems.

the only pose that minimizes both errors is the intersection of the curves in the sets $_\diamond\bar{L}_{\mathsf{L}}$ and $_\diamond\bar{L}_{\mathsf{F}_1}$. Since such curves can have up to 4 intersections, in practice we must assume that the intersection we are interested in is the closest to the observed feature. This is common in IBVS approaches [57], where it is assumed that the error to the desired configuration in the image plane is kept small. Note that agents L and F_1 must be in the correct formation geometry – after applying the method of Section 6.4– for the $M > 2$ agent to converge to the desired relative position.

Remark 3 *It is important to remark that in a formation of three or more agents, only one of these requires a range measurement to the leader. All the other followers benefit from the intersection of epipolar curves $_\diamond\bar{L}_j$ and $_\diamond\bar{L}_m$, $j, m \in M$, to define their relative pose in the formation, solely using image features. Such distance can be obtained with ultra wideband devices, with below-centimeter error [102].*

Let $_\diamond\bar{F}_{\mathsf{LF}_1} = \{_\diamond\bar{f}_1^{\mathsf{LF}_1}, ..., _\diamond\bar{f}_5^{\mathsf{LF}_1}\}$ be the image features corresponding to the intersection of the line sets $_\diamond\bar{L}_{\mathsf{L}}$ and $_\diamond\bar{L}_{\mathsf{F}_1}$. In this case, we wish to minimize the error between $_\circ f_i^{\mathsf{F}_j}$ and $_\circ \bar{f}_i^{\mathsf{LF}_1}, i = 1, \ldots, 5, j = 2, \ldots, \mathsf{M} - 2$, which are features in the image plane. Furthermore, note that we can obtain the non-lifted representation of these points with (2.14), meaning we can minimize the error in the non-lifted coordinate space. Considering the advantages of the Model Predictive Controller scheme proposed in Section 6.4, we define the error state

$$\epsilon_j = \begin{bmatrix} \bar{f}_1^{\mathsf{LF}_1} - f_1^{\mathsf{F}_j} \\ \vdots \\ \bar{f}_5^{\mathsf{LF}_1} - f_5^{\mathsf{F}_j} \end{bmatrix}, \tag{6.22}$$

where $\epsilon_j \in \mathbb{R}^{10} \subset \mathcal{E}_j$ concatenates the image-based feature errors. The set

$$\mathcal{E}_j \triangleq \{ \epsilon_j \in \mathbb{R}^{10} : \bar{f}_i^{\mathsf{LF}_1} - 1_{2\times 1} \leq \epsilon_{j,[i]} \leq 1_{2\times 1} + \bar{f}_i^{\mathsf{LF}_1} \} \tag{6.23}$$

contains all admissable image errors i is associated with the error of each feature: $\epsilon_{j,[i]} := \bar{f}_i^{\mathsf{LF}_1} - f_i^{\mathsf{F}_j}$. Accordingly, the error dynamics for follower j are defined as

$$\dot{\epsilon}_j = \begin{bmatrix} L_1(f_1^{\mathsf{F}_j})^T & \cdots & L_5(f_5^{\mathsf{F}_j})^T \end{bmatrix}^T u^{\mathsf{F}_j}. \tag{6.24}$$

As before, due to the discrete-time nature of the MP-IBFC controller, we discretize (6.24) as

$$\epsilon_j(k+1) = g_\epsilon^f(\epsilon_j(k), u^{\mathsf{F}_j}(k)), \tag{6.25}$$

where g_ϵ^f is a first-order Euler discretization of (6.24) with a sampling time Δt. Note that, for this controller, it is possible to write $\epsilon_j(k) = \bar{\xi}^{\mathsf{F}_j}(k) - \xi^{\mathsf{F}_j}(k)$ and therefore $\xi^{\mathsf{F}_j}(k) = \bar{\xi}^{\mathsf{F}_j}(k) - \epsilon_j(k)$, and thus the error is a bijective function of the state $\xi^{\mathsf{F}_j}(k)$, as defined in (6.4). Finally the controller for followers $\mathsf{F}_j, j = 2, \ldots, \mathsf{M} - 2$ is formally given by

$$\min_{\mathbf{u}^{\mathsf{F}_j}} J(\epsilon_j(k), u^{\mathsf{F}_j}(k)) \tag{6.26a}$$

$$\text{subject to: } \epsilon_j(n+1|k) = g_\epsilon^f(\epsilon_j(n|k), u^{\mathsf{F}_j}(n|k)), \tag{6.26b}$$

$$\epsilon_j(n|k) \in \mathcal{E}_j, \tag{6.26c}$$

$$u^{\mathsf{F}_j}(n|k) \in U \tag{6.26d}$$

$$n = 0, \ldots, N - 1 \tag{6.26e}$$

$$\epsilon_j(0|0) = \epsilon_j(0) \tag{6.26f}$$

With this controller, we may state the following result:

Theorem 4 *Under the control law in (6.26) and the assumptions in Problem (2), (6.24) is asymptotically stablilized to the origin, with $j = 2, \ldots, \mathsf{M} - 2$.*

The proof follows the same steps as the proof of Thm. 3, considering $\tilde{L} = \begin{bmatrix} L_1(f_1^{\mathsf{F}_j})^T & \cdots & L_5(f_5^{\mathsf{F}_j})^T \end{bmatrix}^T$, and is omitted here for brevity.

6.6 Results

In this section, we present results from both simulated and experimental data. The derived Model Predictive Control laws ran on an `Intel Core-i7 CPU`, single-threaded process, with 16GB of Random Access Memory available. The implementation operates through `ACADO` [34], which runs `qpOASES` [103] as a low-level solver, and in CasADi [104], with IPOPT solver [105]. Each solver was implemented considering a sampling time of $\Delta t = 0.1$ seconds, and with a receding horizon length of $N = 15$.

The source code will be made available as a framework for Python in `github.com/Pedro-Roque/viscoor`, but also with a ROS [106] wrapper in `github.com/Pedro-Roque/viscoor_ros`.

Simulated Data

The simulation setup consists of Matlab and Python scripts that implement the controllers of Section 6.4 and 6.5. In particular, in Matlab, we simulate the case of Section 6.4, with two perspective cameras, while in Python we simulate the case of 6.5 with multiple camera types for each agent.

All agents had the same input constraints, given by

$$U^j \triangleq \{u^j \in \mathbb{R}^6 : -\mathbf{0.1} \leq u^j \leq \mathbf{0.1}\}, \ j \in \mathrm{M}, \tag{6.27}$$

where $\mathbf{0.1} = [0.1\ 0.1\ 0.1\ 0.1\ 0.1\ 0.1]^T$ corresponding to a maximum of $0.1[\mathrm{m/s}]$ and $0.1[\mathrm{rad/s}]$, while their state constraints follow (6.23) and

$$\mathcal{E} \triangleq \{\epsilon \in \mathbb{R}^6 : 1_{6 \times 1} \leq \epsilon \leq 1_{6 \times 1}\}, \tag{6.28}$$

defines the error set in (6.9), representing a maximum error of $1[\mathrm{m}]$ in distance and a maximum feature-to-curve distance of 1 normalized units. Arbitrary desired relative positions and attitudes, containing common features in the cameras image, were generated to test the algorithms in simulation.

Two Agent Coordination

We first demonstrate the case of two agent coordination with two perspective cameras, representing the available experimental setup. In Fig. 6.4 we observe the result of the simulation, running in Matlab with ACADO. We observe that the feature error quickly converges to zero, followed by the respective attitude and position errors. We observe that the controller manages to asymptotically stabilize the system without violating the control or state constraints. We emphasize that the simulation setup tests a 6-DoF system, where the camera can freely move considering the 3-DoF linear and angular velocity inputs. Average computational cost was kept below 10[ms].

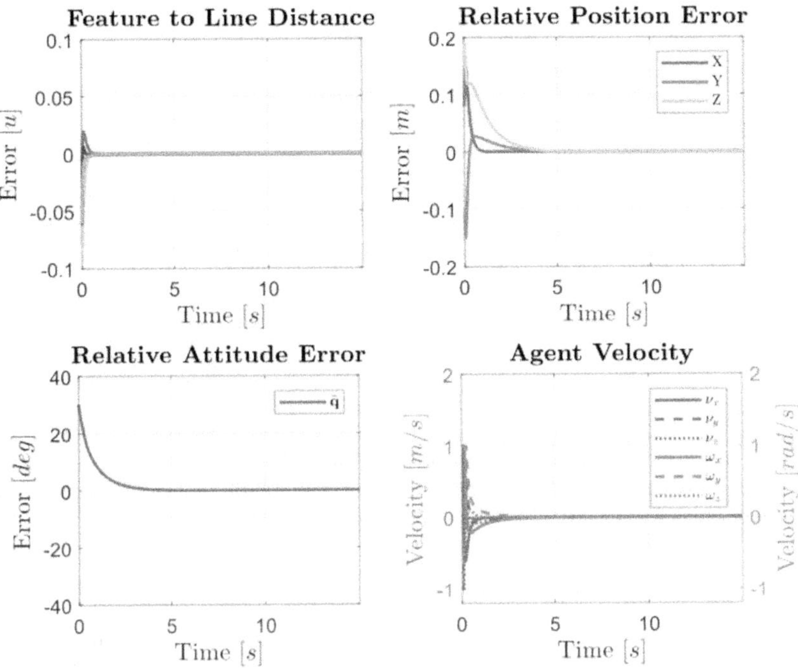

Figure 6.4: Simulation results for two-agent coordination with two perspective cameras. It is possible to observe that the agent asymptotically converges towards the desired state with zero steady-state error on the image-plane, but also on its cartesian position and attitude. Both control and state constraints, given in (6.28) and (6.27) were respected.

Formation Control

On what respected the formation control scenario, we tested in a Python with CasADi a simulation environment with 3 generic camera systems, implementing the approach in Section 6.5. In particular, we tested the case of 2 parabolic cameras (for agents L and F_2) and a *fish-eye* camera (for agent F_1). Figure 6.5 depicts the normalized image plane for the agent F_2, and contains the epipolar curves corresponding to the leader L and neighboring follower F_1 (in blue and purple, respectively), the intersection of these curves (green crosses) and the normalized observed image features (in red). In particular, Fig. 6.5a shows the starting image state, where we can observe an error between the intersection of the epipolar curves and the corresponding image-features. After the regulation task is complete, we observe in Fig. 6.5b that the features are overlaid with the intersection of the

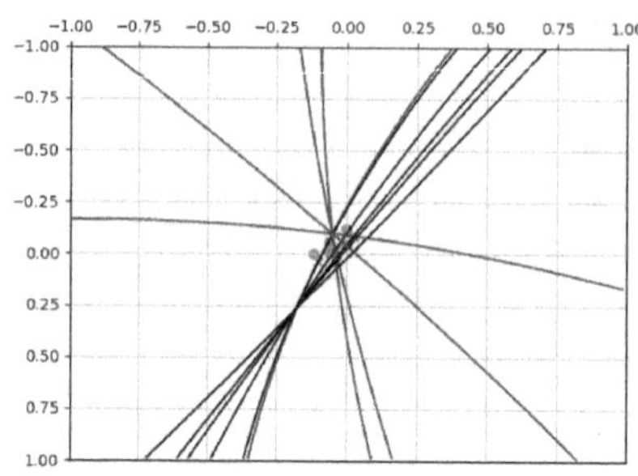

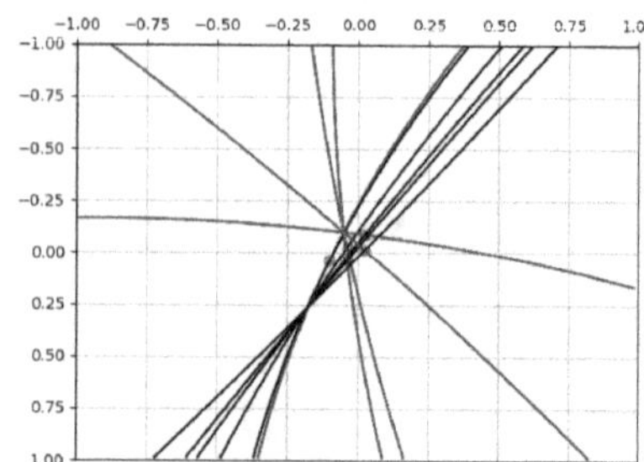

(a) Normalized image plane at the starting configuration.

(b) Normalized image plane at the final configuration.

Figure 6.5: The normalized image plane contains the epipolar curves (blue and purple, for leader L and follower F_1), the intersection of the two sets of epipolar curves (green cross) and the observed features (red blobs). In Fig. 6.5a it is possible to observe the initial configuration for the agent, with an error towards the intersection of the lines. In Fig. 6.5b we observe that the agent successfully converged to the desired configuration.

epipolar curves. The results for this scenario are shown in Fig. 6.6, where we may observe that the feature and attitude error converge to their desired values. in less than 6 seconds. It is however possible to observe that, in the transient behavior, the attitude and position error do not converge uniformly, even though the feature error does. Nonetheless, all input and state constraints were respected. Computational times were also kept under the 10[ms] mark, while in this scenario a discretization of $\Delta t = 0.01$[s] was used.

Experimental Data

Two Agent Coordination

We use a 3 DoF, 4-Wheel Holonomic Drive, Nexus Mecanum robots[2] (shown in Fig. 6.7), modified and interfaced with the Robot Operating System (ROS). Each agent carries an on-board computer (Nvidia Jetson TX2 or Intel NUC with a Core-i7 CPU) with and a Logitech C270 camera. Range measurements were obtained with a Motion Capture System, without loss of generality (obtaining these onboard can be achieved through Ultra-wide Band devices with centimeter accuracy). The on-board computers are used for the detection of five black blobs using the ViSP library [107, 108]. Recalling the definition of the control input for the camera u in

---

[2]https://www.active-robots.com/brands/nexus-robot.html, accessed March 2, 2022.

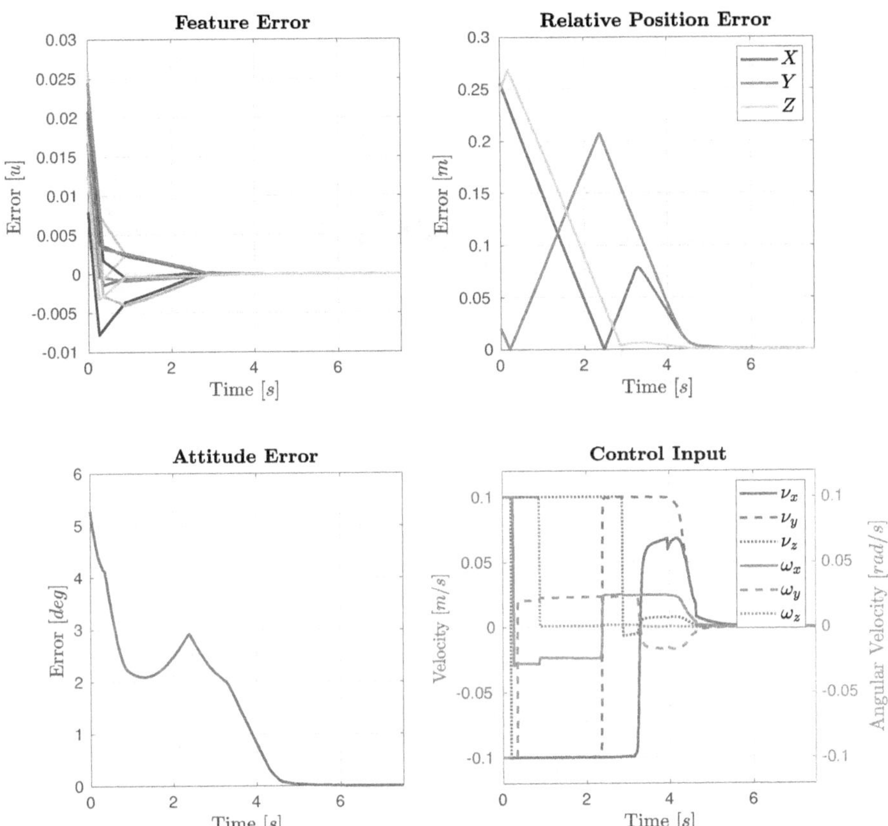

Figure 6.6: Simulation results for the case of 3 agent formation control. At the end of the regulation, we observe that the agent converges to the desired state without residual error. However, it is possible to observe that, in the transient behavior, the attitude and position error do not converge uniformly, even though the feature error does. All state and input constraints were also respected in this task.

Figure 6.7: Experimental setup for the 2-agent coordination scenario. In the picture, we observe two NEXUS Mecanum 4-WD robots observing 5 features in a box. Each agent image-frame is observed on bottom left corner.

(2.25), we set the control limits to

$$U^j \triangleq \{u^j \in R^6 : \|\nu^j_{[1,2]}\| \le 0.08, \|\nu^j_{[3]}\| = 0,$$
$$\|\omega^j_{[1,2]}\| = 0, \|\omega^j_{[3]}\| \le 0.07\}, \tag{6.29}$$

while keeping the state constraints the same as in (6.27). Note that (6.29) also constrains the more general system in (6.10) to three degrees of freedom, for the platform at use.

We observe that the controller correctly converges the system to the desired relative pose with respect to the leader, while respecting the constraints imposed on the state and input during the transient stage. Note that the system will move slowly due to constraints in the platform speed, while yielding an optimal solution for the problem at hand. Due to the simplicity of MPC when incorporating constraints, we achieve the formation goal, while adapting the velocity of the platform such that the features are properly tracked at all times.

Formation Control

For the demonstration of a coordination task involving three agents, we setup an experimental testbed involving two perspective and one *fish-eye* cameras attached to three HEBI Robotics Omni-Directional Mobile Base Kits. Each robot, as in the two-agent scenario, provides a total of 3-DoF.

For this task, instead of tracking black blobs in the environment, we opted to implement a full-stack of feature detection, matching and tracking based on the

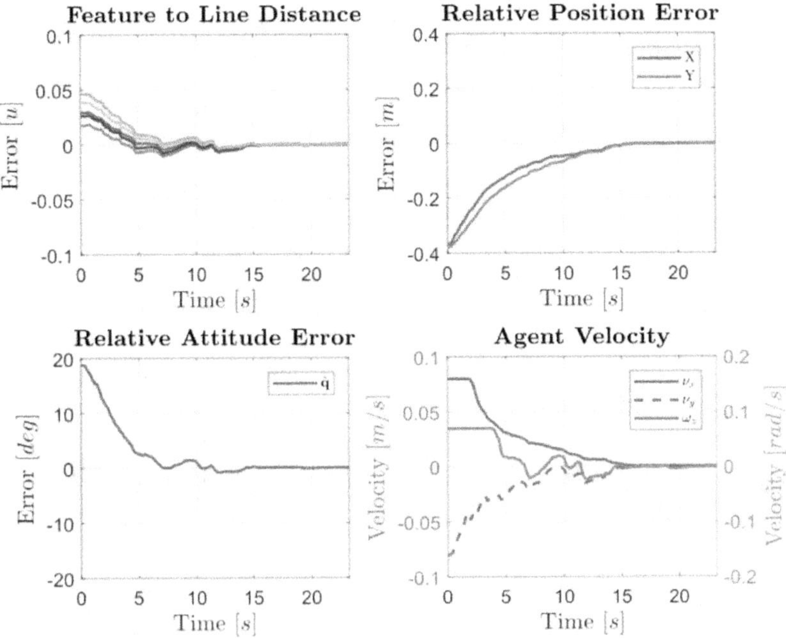

Figure 6.8: Experimental results for the 2-agent controller from Sec. 6.4. The leader agent is stationary, and the follower converges to the desired relative positionWe observe that both the position and attitude error converge to the desired value (and zero error). It is also possible to observe that the control and state constraints are respected during the operation.

OpenCV library [109]. The full source-code will be available to the community at `github.com/Pedro-Roque/viscoor_ros`.

Let us consider the workspace in Fig. 6.9. Each agent in the coordination task runs a ROS node for feature extraction. In this case, we use SIFT features and collect 500 features. Then, F_2 runs a feature matching algorithm based on the FLANN matcher, followed by a RANSAC to obtain the inliers between the perspectives of all the agents. From the set of inliers, we choose 5 features to perform the control task on. Figure 6.10 shows the image-frame of F_2, where it is possible to see the set of all inlier features (black and white circles), as well as the features used for the control task (white circles). If a particular feature that is being tracked is lost, it is automatically replaced by another feature of the inlier set.

Given a set of 5 features, we then calculate the respective epipolar curves and intersections. These can be observed on Fig. 6.11, along with the observed features in red. We also observe that the features converge correctly to the intersection of the two sets of curves, which is associated with a convergence towards the desired formation setting, observed in Fig. 6.12. In this figure, it is also possible to observe a small static error in the image features, associated with the a static control input to the system. We observe that the platform cannot move with such low control inputs, but the associated pose error is small: less than 1[deg] in attitude and 1[cm] in relative position error, for both X and Y - measured in the inertial frame. We also observe an overshoot on ν_x and ω_y, which we concluded to come from a non-smooth motion caused by the holonomic wheels in each agent, which lead to small overshoots in the agent's motion. It is also worth noting that the control input, obtained by solving the FHOC problem in (6.26) is calculated at an average 8[ms], compatible with the frequency at which the control input is applied to the platform, 10[Hz].

It is important to remark that the pipeline here applied can be transferred to a deployment outside laboratory conditions, as the feature detection and matching algorithms can work in unstructured environments.

Figure 6.9: Workspace where the three agent formation operates. Three holonomic platforms are distributed across the workspace, with their cameras observing the same region of interest. Two agents contain perspective cameras, while the third agent (in the center), contains a fisheye lens. The experiments were performed at the Smart Mobility Lab, in KTH.

Figure 6.10: Feature tracking in the image plane of F_2. The robot collects the extracted SIFT features from the neighboring agents, sent over the network, and matches them with the ones collected locally. Then, from the set of common features (represented as black and white circles), five features are selected for tracking (shown in white) using a sparse optical flow algorithm from the OpenCV library.

6.7 Discussion

In this chapter, we derived two control laws using epipolar constraints for multi-agent relative pose coordination: one for coordination among two agents, and another for formation control considering a higher number of agents, observing common points of interest in the world. A stability analysis of the proposed control laws was derived, and both simulation and experimental results demonstrate the success of the proposed approaches in heterogeneous platforms and environments.

In the future, we will investigate the application of such strategies to higher-order nonlinear systems, explore alternative methods to avoid the need for relative distance measurements for the two-agent case, as well as book the tracking of a leader velocity.

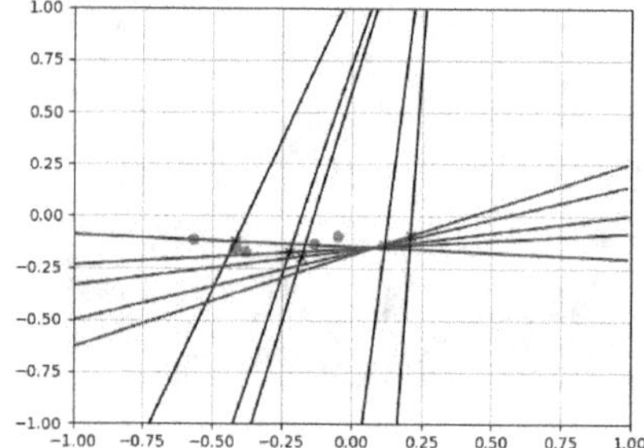

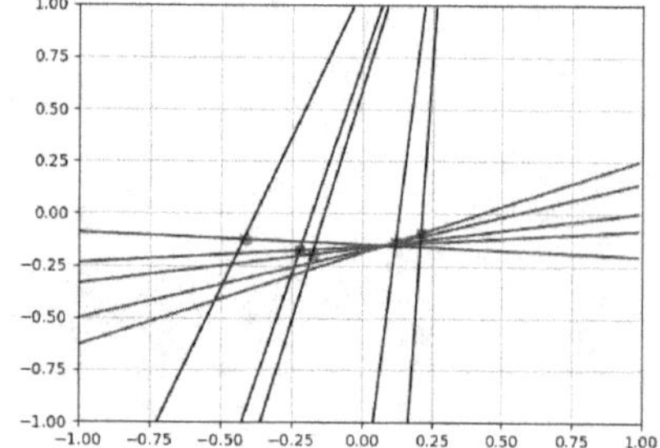

(a) Normalized image plane at the starting configuration.

(b) Normalized image plane at the final configuration.

Figure 6.11: The normalized image plane for the experimental setup, containing the epipolar curves (blue and purple, for leader L and follower F_1), the intersection of the two sets of epipolar curves (green cross) and the observed features (red blobs). It is possible to observe, from Fig. 6.11a to 6.11b, that the agent successfully converges the observed features towards the intersection of the two sets of lines, leading to a convergence towards the correct formation setting with a small static error, as seen in Fig. 6.12.

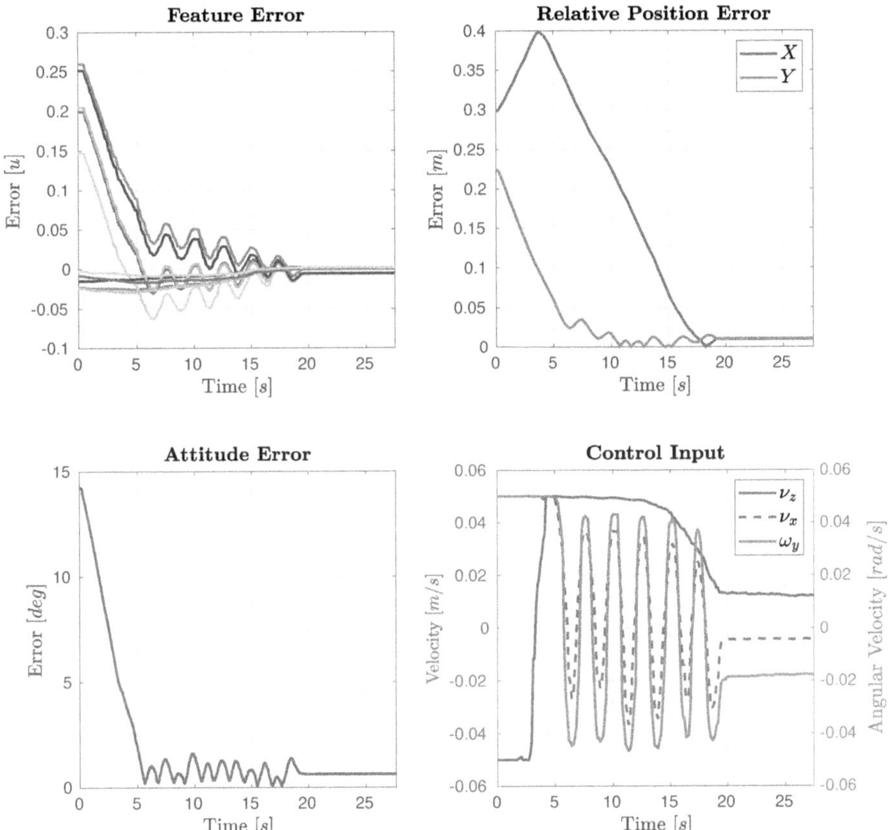

Figure 6.12: Experimental results for the case of three agents formation control. We observe that the agent converges to the correct formation setting with a small steady state error of 1[cm] in position (for the X and Y inertial coordinates) and less than 1[deg] in attitude. It is also possible to observe that the control input suffers an oscillation on ν_x and ω_y, due to the non-smooth motion caused by holonomic wheels. A small steady-state control input is observed, as the platform is not able to move when under small control inputs.

Chapter 7

Corridor Model Predictive Control

The safe control of autonomous systems has received growing attention from the academic community over the past few years. As of recently, the International Space Station became inhabited by autonomous free-flyers that will autonomously perform tasks that can range from inspection to human interaction, where safety - of the autonomus agent, but also of the humans and structures involved - is key. Safety is defined as navigation on a bounded error set in the neighborhood of a possibly time-varying reference. Although safe control is crucial for all autonomous systems that operate in a shared workspace with humans and obstacles, the optimality of such control inputs plays an important long-term role. For instance, optimality affects the operation time of a space-craft, that is linked with the fuel usage or actuator wear and tear. Therefore, safety-critical systems should be endowed with both safe and optimality-aware controllers that respect control input limits and safety constraints, while minimizing control effort.

7.1 Literature Review

MPC and its nonlinear extension (NMPC) [12, 58, 110, 111] are commonly used to address constraint satisfaction and optimality. In short, NMPC is a finite-horizon optimal controller (FHOC) that optimizes a cost function - usually a function of the error and control effort - along a finite time horizon, while enforcing state and control constraints. Extensions of MPC to handle unknown disturbances have been suggested in the literature, namely constraint-tightening MPC [112] and tube-based MPC [113]. On the latter we observe a combined approach of online and offline controllers that might run at different frequencies, where the online component is an ancillary controller. Ancillary controllers rely on associated Robust Control Invariant (RCI) sets [58, Def. 3.6]. In [113] a Monte-Carlo approach is suggested to estimate RCI sets for nonlinear systems, and in [114] the use of a sliding mode controller as the ancillary controller is proposed to exploit the associated RCI set. Another approach to estimate RCI sets has been proposed in [115] where a deep neural network uses Gaussian Regression over several state-space trajectories of the system. With respect to tracking of time-varying references, trajectory tracking methods using linear MPC have been proposed under linear time-varying [116] and linear parameter-varying formulations [117]. With respect to the nonlinear case (NMPC), recent works [118, 119] propose the calculation of a stabilizing terminal cost to show exponential stability towards the reference being tracked, even in the case of dynamic, non-compliant targets. Safety-focused control methods have gained traction in the literature. For instance, Zeroing Control Barrier Functions (ZCBFs) [120] allow a system to operate within safe bounds even when a nominal controller fails, acting as a safety filter that forces the system to evolve inside a safe set that is similar to an RCI set. The robustness properties of such control methods have been inspected in [121, 122]. The authors in [123] have further proposed sampled-data control barrier functions. A combination of ZCBFs with MPC has been proposed in [124] for setpoint stabilization, which however does not provide continuous-time guarantees, as opposed to what we present in this chapter.

In this chapter, we aim at deriving safe control laws with the goal to stay within a desired bound to a given time-varying trajectory while minimizing a cost function over a specified planning horizon. To this end, we take advantage of the recent developments in ZCBFs and NMPC to derive an alternative to existing trajectory tracking controllers, hereafter referred to as Corridor MPC (CMPC). This approach allows for safe trajectory tracking under the optimal control inputs derived using NMPC, while providing safety guarantees for continuous-time systems. Moreover, we present stability and recursive feasibility guarantees for CMPC. In summary, the motivation and advantages of this alternative approach over the state-of-the-art are (i) a sampled-data ZCBF formulation that ensures robust safety of a time-varying set that can be encoded into an MPC framework; (ii) continuous-time safety guarantees in the presence of uncertainties; (iii) an alternative formulation paving the way for further investigation of similarities and differences to existing RCI tube methods.

7.2 Error Dynamics

Let us consider the disturbed dynamics in (2.1). We define the error of the state $\tilde{x}(t)$ with respect to a possibly time-varying trajectory $x_r(t)$ as

$$\tilde{e}(t) = \tilde{x}(t) - x_r(t), \tag{7.1}$$

with the dynamics of the error $\tilde{e}(t)$

$$\dot{\tilde{e}} = f_{c,\tilde{x}}(\tilde{e}(t) + x_r(t), u(t), w(t)) - \dot{x}_r(t). \tag{7.2}$$

Assumption 3 *The reference trajectory $x_r(t)$ satisfies*

$$x_r((k+1)\Delta t) = f(x_r(k\Delta t)) + g(x_r(k\Delta t))u_r(k\Delta t)$$

at the sampling times $k\Delta t$, for $x_r(k\Delta t) \in \mathbb{X}$ and $u_r(k\Delta t) \in \mathbb{U}_r \subset \mathbb{U}$, and is continuously differentiable for all $t \in [k\Delta t, (k+1)\Delta t]$.

Remark 4 *The upper bound on $\mathbb{U}_r$ is related to the upper bound on $\|\dot{x}_r(t)\|$ and $\|w(t)\|, \forall t \in \mathbb{R}_{\geq 0}$. We elaborate on this relation in Section 7.6.*

In discrete-time, the disturbed and nominal system errors are represented as

$$\tilde{e}(k\Delta t) = \tilde{x}(k\Delta t) - x_r(k\Delta t) \text{ and} \tag{7.3}$$

$$e(k\Delta t) = x(k\Delta t) - x_r(k\Delta t), \tag{7.4}$$

whereas their respective dynamics satisfy

$$\tilde{e}((k+1)\Delta t) = f_{\tilde{e}}(\tilde{e}, u, w) =$$
$$= f_{\tilde{x}}(\tilde{e}(k\Delta t) + x_r(k\Delta t), u(k\Delta t), w(k\Delta t)) - \dot{x}_r(k\Delta t), \tag{7.5a}$$
$$e((k+1)\Delta t) = f_e(\tilde{e}, u) =$$
$$= f_x(e(k\Delta t) + x_r(k\Delta t), u(k\Delta t), 0) - \dot{x}_r(k\Delta t), \tag{7.5b}$$

and $f_e(\tilde{e}, u) \triangleq f_{\tilde{e}}(\tilde{e}, u, 0)$. We denote by $\Psi_{\tilde{e}}(k\Delta t, \tilde{e}(0), \mathbf{w}_k)$ the solution of $f_{\tilde{e}}(\tilde{e}, w)$ (where $f_{\tilde{e}}(\tilde{e}, w)$ is $f_{\tilde{e}}(\tilde{e}, u, w)$ given a feedback controller $u(k\Delta t) = K(\tilde{e}(k\Delta t))$) at time $k\Delta t$, with initial condition $\tilde{e}(0)$ and given a sequence of disturbances $\mathbf{w}_k = \{w(0), w(\Delta t), ..., w(k\Delta t)\}, \forall k \in \mathbb{N}_0$. We denote the solution of the nominal system as $\Psi_e(k\Delta t, \tilde{e}(0)) \triangleq \Psi_{\tilde{e}}(k\Delta t, \tilde{e}(0), \mathbf{0}_k)$.

7.3 Problem Statement

In this work, we will focus on the reference tracking problem for Nonlinear Model Predictive Control. In particular, we are interested in robustly following a time-varying reference, that is, to stay within an ε-small neighborhood of the desired reference, even in the presence of an additive noise $w(t) \in \mathbb{W}$. Moreover, as we plan to apply our solution in safety-critical systems, it is required that the state $\tilde{x}$ remains in the predefined safe set $\mathcal{C}(t)$, ε-close to the desired reference trajectory $x_r(t)$, and that the control input $u(k\Delta t)$ satisfies the actuation limits set by $u(k\Delta t) \in \mathbb{U}$.

Formally, we define the problem as follows:

Problem 3 *Let the dynamics for $\tilde{x}(t)$ be defined by (2.1) and let $\tilde{e}(t)$ be as in (7.1). Let the admissible safe set $\mathcal{C}(t)$ be defined by (2.6) for $h(\tilde{x}, t) = \varepsilon^2 - \|\tilde{x} - x_r(t)\|_2^2$, where $\varepsilon \in \mathbb{R}_{>0}$ is the maximum allowed error to the reference and $x_r(t)$ is a time-varying reference, as in Assumption 3. The goal is to design a feedback control law $K_N(\tilde{e}(k\Delta t)) \in \mathbb{U}, k \in \mathbb{N}_0$ through (2.5), such that $\tilde{x}(t)$ remains in $\mathcal{C}(t)$ – and therefore $\|\tilde{e}(t)\| \leq \varepsilon, \forall t \geq 0$, – while minimizing a cost function $J(e(k\Delta t), u(k\Delta t))$, where $e(k\Delta t)$ is as in (7.4).*

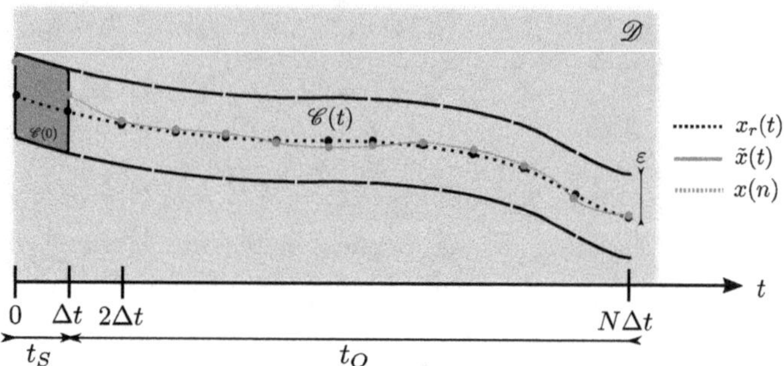

Figure 7.1: Corridor MPC prediction horizon detail. From time $t_s \in [0, \Delta t]$, using an optimally safe control input, the state $\tilde{x}$ is ensured to evolve within the safe set $\mathcal{C}(t)$, defined with ZCBFs h, and ε maximum deviation from $x_r(t)$. The safe control input applied at $t = 0$ is optimized taking into account the optimality of the solution for the entire receding horizon planning, $t \in \{t_S \cup t_O\}$. The procedure is repeated at all sampling times, ensuring that $\tilde{x} \in \mathcal{C}(t)$ through the whole operation.

7.4   Corridor MPC

To solve Problem 1 we propose CMPC, a novel approach that takes advantage of the robust and safety properties of ZCBFs and the constrained optimality of NMPC.

ZCBF for Sampled-Data Systems

ZCBFs guarantees, as presented in the literature (e.g. in [13]), are targeted at continuous-time systems and, as argued before, require locally Lipschitz continuous control laws $u(x, t)$. As in this work we integrate ZCBFs into a NMPC that operates at discrete time steps $k\Delta t, k \in \mathbb{N}_0$, we instead use sampled-data control barrier functions as proposed in [123]. To this end, robustness margins can be incorporated in the ZCBF framework (2.7), ensuring that the system remains safe in-between sample times. The formulation proposed in [123] ensures continuous-time safety using a sampled-data control barrier function, assuming that the constraint set $\mathcal{C}$ is compact and the system dynamics are bounded. Lemma 2 extends the result in [123, Lem. 4] for a time-varying safe set $\mathcal{C}(t)$.

Definition 2 (SD-ZCBF) *Consider the system (2.1) and the compact sets $\mathcal{C}(t)$ and $\mathcal{D}$, where $\mathcal{C}(t)$ is defined by (2.6) for a continuously differentiable function $h : \mathcal{D} \times \mathbb{R}_{\geq 0} \to \mathbb{R}$, and $\mathcal{D} \supset \mathcal{C}(t)$ for all $t \geq 0$. The function h is a sampled-data zeroing control barrier function (SD-ZCBF) if for a given $\Delta t > 0$ there exists an extended class-$\mathcal{K}$ function α, where $\alpha \circ h : \mathcal{D} \times \mathbb{R}_{\geq 0} \to \mathbb{R}$ is Lipschitz continuous on $\mathcal{D} \times \mathbb{R}_{\geq 0}$, such that for any point $\tilde{x} \in \mathcal{D}$ and $k \in \mathbb{N}_0$ there is a constant feedback* *input $u(\tilde{x}, k\Delta t) \in \mathbb{U}$, shortened to $u(k\Delta t)$, satisfying the following condition:*

$$L_{f_c}h(\tilde{x}, k\Delta t) + L_{g_c}h(\tilde{x}, k\Delta t)u(k\Delta t)$$
$$+ \alpha(h(\tilde{x}, k\Delta t)) \geq \bar{\nu} + c_6 + c_7\bar{w}, \tag{7.6}$$

with $\bar{\nu} \geq \nu(\Delta t)$, where

$$\nu(\Delta t) := ((\bar{c}_1 + \bar{c}_2 + \bar{c}_3 c_4)(c_5 + \bar{w}) + (\underline{c}_1 + \underline{c}_2 + \underline{c}_3 c_4))\,\Delta t, \tag{7.7}$$

and where $\bar{c}_1, \bar{c}_2, \bar{c}_3 \in \mathbb{R}_{>0}$ are the respective Lipschitz constants of $\alpha \circ h$, $L_{f_c}h$, $L_{g_c}h$ with respect to $\tilde{x}$, $\underline{c}_1, \underline{c}_2, \underline{c}_3 \in \mathbb{R}_{>0}$ are the respective Lipschitz constants of $\alpha \circ h$, $L_{f_c}h$, $L_{g_c}h$ with respect to t, and $c_4, c_5, c_6, c_7, \bar{w} \in \mathbb{R}_{>0}$ are the respective uniform bounds of $\|u\|$, the dynamics $\|f_c(\tilde{x}) + g_c(\tilde{x})u\|$, $\|\frac{\partial h(\tilde{x}(k),t)}{\partial t}\|$, $\|\frac{\partial h(\tilde{x}(k),t)}{\partial \tilde{x}}\|$, and $\|w\|$ on $\mathcal{D} \times \mathbb{R}_{\geq 0}$.

The intuition behind Definition 2 is given by Lemma 2.

Lemma 2 *For a given $\Delta t > 0$, suppose $h : \mathcal{D} \times \mathbb{R}_{\geq 0} \to \mathbb{R}$ is a SD-ZCBF for the system (2.1). For any $k \in \mathbb{N}_0$, let $t_0 = k\Delta t$ and $\tilde{x}(t_0) = \tilde{x}(k\Delta t) \in \mathcal{C}(t_0) \subset \mathcal{D}$. Consider (2.1) in closed-loop with $u(t) = u(\tilde{x}(k\Delta t), k\Delta t) \in \mathbb{U}\ \forall t \in [k\Delta t, (k+1)\Delta t]$, for which (7.6) holds, on $[k\Delta t, (k+1)\Delta t]$. Then $\tilde{x}(t)$ is uniquely defined on $[k\Delta t, (k+1)\Delta t]$, and $\tilde{x}(t) \in \mathcal{C}(t)$ for all $t \in [k\Delta t, (k+1)\Delta t]$.*

Proof: Since the dynamics of (2.1) are locally Lipschitz continuous on $\mathbb{R}_{\geq 0} \times \tilde{\mathbb{X}}$, piece-wise continuous in t, and u is constant on $[k\Delta t, (k+1)\Delta t]$, then $\tilde{x}(t)$ is uniquely defined on $[k\Delta t, \tau_1]$ for some $\tau_1 > k\Delta t$ [125, Thm. 54]. We note that if $\tau_1 \geq (k+1)\Delta t$, then of course $\tilde{x}(t)$ is uniquely defined on $[k\Delta t, (k+1)\Delta t]$. Suppose then that $\tau_1 < (k+1)\Delta t$. Furthermore, since $\tilde{x}(k\Delta t) \in \mathcal{C}(k\Delta t) \subset \mathcal{D}$, then by continuity of $\tilde{x}$ there exists a $\tau_0 \in (k\Delta t, \tau_1]$ for which $\tilde{x}(t) \in \mathcal{D}$ for all $t \in [k\Delta t, \tau_0]$. We define the following function on $[k\Delta t, \tau_0]$:

$$m_k(t) = \left(L_{f_c}h(\tilde{x}(t), t) - L_{f_c}h(\tilde{x}(k\Delta t), k\Delta t)\right)$$
$$+ \left(\alpha(h(\tilde{x}(t)), t) - \alpha(h(\tilde{x}(k\Delta t), k\Delta t))\right)$$
$$+ \left(L_{g_c}h(\tilde{x}(t), t) - L_{g_c}h(\tilde{x}(k\Delta t), k\Delta t)\right)u(k\Delta t). \tag{7.8}$$

By the Lipschitz properties of $f_c(\tilde{x})$, $g_c(\tilde{x})$, $h(\tilde{x}, t)$, $\alpha(h(\tilde{x}, t))$, and the fact that $k\Delta t < \tau_0 < (k+1)\Delta t \Rightarrow \tau_0 - k\Delta t < \Delta t$, it follows that:

$$\|m_k(t)\| \leq (\bar{c}_1 + \bar{c}_2 + \bar{c}_3\|u(k\Delta t)\|)\|\tilde{x}(t) - \tilde{x}(k\Delta t)\|$$
$$+ (\underline{c}_1 + \underline{c}_2 + \underline{c}_3\|u(k\Delta t)\|)\Delta t. \tag{7.9}$$

Next we introduce the bounds on the system dynamics $\|f_c(\tilde{x}) + g_c(\tilde{x})u\|$, denoted $c_5 \in \mathbb{R}_{>0}$, and on the input u, denoted $c_4 \in \mathbb{R}$. Note that c_5 is known to exist since the system dynamics are continuous on the compact set $\mathcal{D}$. Furthermore, since $\mathbb{U}$ is compact there exists a c_4 such that $\mathbb{U} \subset \{u \in \mathbb{R}^m : \|u\| \leq c_4\}$. Now, we get $\|\tilde{x}(t) - \tilde{x}(k\Delta t)\| = \int_{k\Delta t}^{\tau_0} \|f_c(\tilde{x}(t)) + g_c(\tilde{x}(t))u(k\Delta t) + w(t)\| dt \leq (c_5 + \bar{w})\Delta t$. Substitution into (7.9) yields: $\|m_k(t)\| \leq \nu(\Delta t)$.

Now we have derived a bound on $\|m_k(t)\|$ for $t \in [k\Delta t, \tau_0]$. We add $L_{f_c} h(\tilde{x}, t) + L_{g_c} h(\tilde{x}, t)u(k\Delta t) + L_w h(\tilde{x}, t) + \frac{\partial h(\tilde{x}, t)}{\partial t} + \alpha(h(\tilde{x}, t))$ to both sides of (7.6) and substitute (7.8) which yields:

$$L_{f_c} h(\tilde{x}, t) + L_{g_c} h(\tilde{x}, t)u(k\Delta t) + L_w h(\tilde{x}, t) + \frac{\partial h(\tilde{x}, t)}{\partial t}$$

$$+ \alpha(h(\tilde{x}, t)) \geq L_w h(\tilde{x}, t) + \frac{\partial h(\tilde{x}, t)}{\partial t} + \bar{\nu} + m_k(t)$$

$$+ c_7 \bar{w} + c_6,$$

Since $0 \leq \|m_k(t)\| \leq \nu(\Delta t) \leq \bar{\nu}$, $\|L_w h\| = \|\frac{\partial h(\tilde{x}, t)}{\partial \tilde{x}} w(t)\| \leq c_7 \bar{w}$, and $\|\frac{\partial h(\tilde{x}, t)}{\partial t}\| \leq c_6$ it follows that $L_{f_c} h(\tilde{x}, t) + L_{g_c} h(\tilde{x}, t)u(k\Delta t) + L_w h(\tilde{x}, t) + \frac{\partial h(\tilde{x}, t)}{\partial t} + \alpha(h(\tilde{x}, t)) \geq 0$ for all $t \in [k\Delta t, \tau_0]$. It is clear then that $\dot{h}(\tilde{x}, t) \geq -\alpha(h(t))$, which by Lemma 2 of [126] ensures that $\tilde{x}(t) \in \mathcal{C}(t)$ for all $t \in [k\Delta t, \tau_0]$. Now we prove that $\tilde{x}(t) \in \mathcal{C}(t)$ for all $t \in [k\Delta t, \tau_1]$ by contradiction. Suppose instead that for some $\tau_a \in (\tau_0, \tau_1]$, $\tilde{x}(\tau_a) \in \mathcal{D} \setminus \mathcal{C}(t)$ and $\tilde{x}(t) \in \mathcal{D}$ for all $t \in [k\Delta t, \tau_a]$ (i.e., the solution has left $\mathcal{C}(t)$, but not $\mathcal{D}$). Then $\tilde{x}(t)$ must leave $\mathcal{C}(t)$ at some $t < \tau_a$. Furthermore, since the closed-loop dynamics are locally Lipschitz on $\mathcal{D}$, $\tilde{x}(t)$ is uniquely defined on $[k\Delta t, \tau_a]$ (this is shown by repeatedly applying [125, Thm. 54] since $\tilde{x}(t)$ remains in $\mathcal{D}$ over which local Lipschitz continuity of the closed-loop dynamics holds). To leave $\mathcal{C}(t)$, $\dot{h}(\tilde{x}, t) < 0$ must hold on $\partial \mathcal{C}(t)$ (2.6b). We recompute $m_k(t)$ by repeating the previous steps and replacing τ_0 with τ_a. We note that since all the previously defined bounds $c_1, ..., c_7, \bar{w}$ are independent of time and $\tau_a \leq (k+1)\Delta t$, that the repeated analysis yields the exact same bound on $m_k(t)$ as in (7.9). Therefore we see that $\dot{h}(\tilde{x}, t) \geq 0$ holds for any $\tilde{x}(t) \in \mathcal{C}(t), t \in [k\Delta t, \tau_a]$. We arrive at a contradiction, and so $\tilde{x}(t)$ can never leave $\mathcal{C}(t)$ (and $\mathcal{D}$) on $t \in [k\Delta t, \tau_1]$. If $\tau_1 < (k+1)\Delta t$, then the solution $\tilde{x}(t)$ must have left every compact subset over which local Lipschitz continuity of the closed-loop dynamics holds (see proof of [127, Thm. 3.3]). However we have just shown that $\tilde{x}(t) \in \mathcal{C}(t) \subset \mathcal{D}$, and the closed-loop system is locally Lipschitz continuous on $\mathcal{D}$. Thus in fact $\tau_1 \geq (k+1)\Delta t$ and so $\tilde{x}(t)$ is uniquely defined on $[k\Delta t, (k+1)\Delta t]$. Finally, we once more compute $m_k(t)$ by repeating the previous steps and substituting τ_0 with $(k+1)\Delta t$. Once again we arrive at the same bound on $m_k(t)$ from (7.9) such that $\dot{h}(\tilde{x}, t) \geq 0$ for all $t \in [k\Delta t, (k+1)\Delta t]$, which ensures that $\tilde{x}(t) \in \mathcal{C}(t)$ for all $t \in [k\Delta t, (k+1)\Delta t]$. $\qquad\square$

Remark 5 *If $u = u(\tilde{x}(k\Delta t), k\Delta t)$ for almost all $t \in [k\Delta t, (k+1)\Delta t]$ and $u \in \mathbb{U}$, then the results of Lemma 2 still hold since conditions of Theorem 54 of [125] and Lemma 2 of [126] hold, and u is bounded. Thus Lemma 2 ensures forward invariance of $\mathcal{C}(t)$ for (2.4) in closed-loop with a zero-order hold control input u satisfying (7.6) at each sampling time.*

Corridor MPC Formulation

Formally, the Corridor MPC, solved at each sampling time $k\Delta t, k \in \mathbb{N}_0$, is formulated as

$$J_N^*(\tilde{e}(k\Delta t)) = \min_{\mathbf{u}_k^*} \quad J_N(e, u) \tag{7.10a}$$

$$\text{s.t.:} \quad x((m+1)\Delta t|k\Delta t) = f_x(x, u) \tag{7.10b}$$

$$L_{f_c}h(x(0|k\Delta t), k\Delta t)$$
$$+ L_{g_c}h(x(0|k\Delta t), k\Delta t)u(0|k\Delta t))$$
$$+ \alpha(h(x(0|k\Delta t), k\Delta t)) \geq \bar{\nu} + c_6 + c_7\bar{w} \tag{7.10c}$$

$$u(m\Delta t|k\Delta t) \in \mathbb{U}, \quad \forall m \in \mathbb{N}_{[0, N-1]} \tag{7.10d}$$

$$e(n\Delta t|k\Delta t) = x(n\Delta t|k\Delta t) - x_r(n\Delta t|k\Delta t) \tag{7.10e}$$

$$x(0|k\Delta t) = \tilde{x}(k\Delta t), \quad \forall n \in \mathbb{N}_{[0, N]} \tag{7.10f}$$

where

$$J_N(e, u) = \sum_{n=0}^{N-1} l(e(n\Delta t|k\Delta t), u(n\Delta t|k\Delta t))$$
$$+ V(e(N\Delta t|k\Delta t)). \tag{7.11}$$

The FHOC yields a feedback optimal control input $K_N(\tilde{e}) = u^*(0|k\Delta t)$, with respect to the cost function $J_N(e, u)$ and subject to the constraints (7.10b)-(7.10f), yielding predicted trajectories for the nominal state $\mathbf{x}_k^*$, and with (7.10e), $\mathbf{e}_k^*$, and control inputs $\mathbf{u}_k^*$, similarly to (2.5).

In this approach, the optimization horizon of the MPC, and consequent control inputs, can be divided in two time phases: the set $t_S := \{t \in \mathbb{R}_{\geq 0} : k\Delta t \leq t \leq (k+1)\Delta t\}$, and the set $t_O := \{t \in \mathbb{R}_{\geq 0} : (k+1)\Delta t < t \leq (k+N)\Delta t\}$, as illustrated in Fig. 7.1. This virtual division is owed to condition (7.10c), which is imposed only for the first-step of the receding horizon. Condition (7.10c) implies (as later proved in Lemma 3) that $\tilde{x}(t) \in \mathcal{C}(t), \forall t \in [k\Delta t, (k+1)\Delta t]$, meeting the safety goals defined in Problem 3. On what concerns the second step of the receding horizon, associated with the prediction time t_O, the FHOC (7.10) is only constrained by the dynamics (7.10b) and control constraints (7.10d) where the goal is to minimize the error $e(k\Delta t)$ and control effort $u(k\Delta t), k \in \mathbb{N}$, subject to (7.10b). When recursively solving the FHOC (7.10) at each sampling time, safety of the continuous time system (2.1) through the whole operation is ensured. Consequently, it is ensured that all control inputs sent to the system are optimally safe.

7.5 Feasibility and Stability Analysis

In this section we present the feasibility and stability analysis derived for the CMPC problem (7.10).

Recursive Feasibility

Lemma 3 details the recursive feasibility of the proposed Corridor MPC defined in (7.10).

Lemma 3 *Let $h : \mathcal{D} \times \mathbb{R}_{\geq 0} \to \mathbb{R}$ be a SD-ZCBF for the system (2.1) and let $\tilde{x}(0) \in \mathcal{C}(0)$. Then the system (2.1) under the control $u^*(\tilde{x}, k\Delta t)$ ensures that the FHOC problem (7.10) is recursively feasible.*

Proof: Since h is a SD-ZCBF as per Definition 2, there exists a $u \in \mathbb{U}$ such that (7.10c) holds for any $\tilde{x}(0) \in \mathcal{C}(0)$. Let $t_0 = 0$ and $t_1 = \Delta t$. At t_0, and using Lemma 2, the control input $u(t_0)$ obtained by solving (7.10) ensures that $\tilde{x}(t) \in \mathcal{C}(t), \forall t \in [t_0, t_1]$. Repeating the same procedure for $t_0 = k\Delta t$ and $t_1 = (k+1)\Delta t, k = 1, \ldots, \infty$, concludes the proof. We note that at $t = (k+1)\Delta t$ a switching of the control input occurs from $u(k\Delta t)$ to $u((k+1)\Delta t)$. However, on the interval $[k\Delta t, (k+1)\Delta t]$, this switch occurs on a set of measure zero such that $u(k\Delta t)$ is applied for almost all $t \in [k\Delta t, (k+1)\Delta t]$. Now, from Remark 5, it is clear that the results of Lemma 2 still hold. Then $\qquad\square$

Input-to-State Stability

Input-to-State Stability (ISS) serves as an adequate framework to obtain stability bounds when controlling the system in (2.3).

Assumption 4 *There exists a $h_e(\tilde{e}(t))$ such that $\forall t \geq 0, \tilde{x} \in \mathcal{D}, h_e(\tilde{e}(t)) = h(\tilde{x}, t)$.*

A common example that satisfies Assumption 4 is $h(\tilde{x}, t) = \varepsilon - \|\tilde{x} - \tilde{x}_r(t)\|_2 = \varepsilon - \|\tilde{e}(t)\|_2 = h_e(\tilde{e}(t))$. Analogously to $\tilde{x}$, let $\mathcal{C}_e(t)$ be defined according to

$$\mathcal{C}_e(t) = \{\tilde{e} \in \mathbb{R}^n : h_e(\tilde{e}(t)) \geq 0\}. \tag{7.12}$$

We are interested on showing that (7.5a) is ISS w.r.t $w(k\Delta t) \in \mathbb{W}$ for all $\tilde{e}(k\Delta t) \in \mathcal{C}_e, k \in \mathbb{N}_0$, as the reference trajectory $x_r(t)$ can be designed respecting this condition. The derivations here presented follow closely the results in [128, Sec. 4], adapted to the dynamics in (7.5). Let ISS be defined according to:

Definition 3 (Input-to-State Stability (ISS) [128]) *Consider that (7.5a) is controlled by a feedback control law $u(k\Delta t) = K_N(e(k\Delta t))$. Then, (7.5a) is ISS w.r.t. $w(k\Delta t) \in \mathbb{W}, k \in \mathbb{N}_0$, if there exist a $\mathcal{KL}$-function β and a $\mathcal{K}$-function γ such that for all initial states $\tilde{e}(0)$ and disturbances $w(k\Delta t) \in \mathbb{W}$,*

$$\|\Psi_{\tilde{e}}(k\Delta t, \tilde{e}(0), \mathbf{w}_k)\| \leq \beta(\|\tilde{e}(0)\|, k\Delta t) + \gamma(\|\mathbf{w}_{k-1}\|), k \in \mathbb{N}. \tag{7.13}$$

For simplicity, let $a \triangleq a(k\Delta t)$, for any function a in the reminder of this section, except where otherwise specified.

Assumption 5 *$V(e)$ is a control Lyapunov function (CLF) for system (7.5b) such that for all $\tilde{e} \in \mathcal{C}_e$ there exist two $\mathcal{K}_\infty$-functions α_V and β_V satisfying $\alpha_V(\|\tilde{e}\|) \leq V(\|\tilde{e}\|) \leq \beta_V(\|\tilde{e}\|)$ and*

$$\min_u \{V(f_e(\tilde{e}, u)) - V(\tilde{e}) + l(\tilde{e}, u)\} \leq 0 \tag{7.14}$$

such that $\tilde{e} \in \mathcal{C}_e, u \in \mathbb{U}$ and $f_e(\tilde{e}, u) \in \mathcal{C}_e$. Moreover, the cost function $J_N(\tilde{e}, \mathbf{u}_k)$ is uniformly continuous such that

$$\|J_N(\tilde{e}_1, \mathbf{u}_k) - J_N(\tilde{e}_2, \mathbf{u}_k)\| \leq \alpha_{J_N}(\|\tilde{e}_1 - \tilde{e}_2\|) \tag{7.15}$$

for $\tilde{e}_1, \tilde{e}_2 \in \mathcal{C}_e$, α_{J_N} being a $\mathcal{K}_\infty$-function and for any control sequence $\mathbf{u}_k \in \mathbb{U}$.

It is known [128] that Assumption 5 is sufficient to prove that the optimal cost $J_N^*(\tilde{e})$ is a Lyapunov function of the closed-loop nominal system, for all $\tilde{e} \in \mathcal{C}_e$, yielding asymptotic stabilization of (7.5b). However, for the disturbed system, we can at best expect to converge to a neighborhood of the origin. Consider now the dynamic system in (7.5a) satisfying Assumption 3, controlled by $u = K_N(\tilde{e})$ which is solution to the FHOC problem (7.10) where the terminal cost $V(\tilde{e})$ respects Assumption 5. In this case, it is shown in [128, Thm. 4] that (7.5a) fulfils the ISS property in $\mathcal{C}_e$ if (i) $f_e(\tilde{e}, u)$ is uniformly continuous for all $\tilde{e} \in \mathcal{C}_e$, or (ii) the optimal cost $J_N^*(\tilde{e})$ is uniformly continuous in $\mathcal{C}_e$. In Theorem 5 it is proven that the latter case can be fulfilled by the CMPC.

Theorem 5 *Consider the system (7.5a) fulfilling Assumption 3, $h(\tilde{x}, t)$ a valid SD-ZCBF and let $h_e(\tilde{e}(t))$ be such that Assumption 2 holds. Let $K_N(\tilde{e})$ be the FHOC in (7.10), where $V(\tilde{e})$ satisfies Assumption 5, for all $\tilde{e} \in \mathcal{C}_e$. Then, system (7.5a) fulfils the ISS property, defined in Definition 3, in the set $\mathcal{C}_e$ for a sufficiently small bound on the uncertainty w.*

Proof: From [128, Prop. 1-C2] we can show that the optimal cost $J_N^*(\tilde{e})$ is uniformly continuous in the set $\mathcal{C}_e$. For any $\tilde{e} \in \mathcal{C}_e$ and $\mathbf{e}_k^*$ (the optimal trajectory obtained by solving the FHOC in (7.10) with initial state $\tilde{e}$), Bellman's optimality principle implies that $J_N^*(\tilde{e}(k\Delta t)) > J_{N-j}^*(e^*(j\Delta t|k\Delta t)), \forall j = 0, 1, ..., N, \forall k \in \mathbb{N}_0$, given that Assumptions 3 and 5 are respected. Then, for all $\tilde{e} \in \mathcal{C}_e$, all predicted optimal trajectories $\mathbf{e}_k^*$ remain in $\mathcal{C}_e$. Let $\tilde{\mathbf{u}}_k$ be a sequence of $N - 1$ control inputs $\tilde{\mathbf{u}}_k = \{u(k\Delta t), .., u((k + N - 1)\Delta t)\}$ obtained by solving the FHOC in (7.10) at the sampling times $t = k\Delta t, ..., (k + N - 1)\Delta t$. In other words, $\tilde{\mathbf{u}}_k = \{u^*(0|k\Delta t), ..., u^*(0|(k + N - 1)\Delta t)\}$. Such control input trajectory $\tilde{\mathbf{u}}_k$ is known to exist since for any $\tilde{e}(k\Delta t) \in \mathcal{C}_e$ there exists a control input $u \in \mathbb{U}$ such that $\tilde{e}((k + 1)\Delta t) \in \mathcal{C}_e$, as long as Assumption 2 holds and $h(\tilde{x}, t)$ is a valid SD-ZCBF. Then, let $\tilde{e}_1, \tilde{e}_2 \in \mathcal{C}_e$ such that $J_N^*(\tilde{e}_1) \geq J_N^*(\tilde{e}_2)$, for which it holds

$$\|J_N^*(\tilde{e}_1) - J_N^*(\tilde{e}_2)\| \leq \|J_N(\tilde{e}_1, \tilde{\mathbf{u}}_k) - J_N(\tilde{e}_2, \tilde{\mathbf{u}}_k)\|$$

$$\overset{(7.15)}{\leq} \alpha_{J_N}(\|\tilde{e}_1 - \tilde{e}_2\|). \tag{7.16}$$

Note that $\|J_N^*(\tilde{e}_1) - J_N^*(\tilde{e}_2)\| \leq \|J_N(\tilde{e}_1, \tilde{\mathbf{u}}_k) - J_N(\tilde{e}_2, \tilde{\mathbf{u}}_k)\|$ since $\|u^*(1|k\Delta t)\| \leq \|u^*(0|(k+1)\Delta t)\|$ due to the ZCBF condition in (7.10c). Therefore, the optimal cost is uniformly continuous in $\mathbb{R}^n$. From [128, Thm. 4], we have that the system (7.5a) is ISS w.r.t. the disturbance $w \in \mathbb{W}$, completing the proof. $\qquad\square$

7.6 Corridor MPC for Free-flyer Kinematics

In this section we apply the CMPC framework to a free-flyer kinematics model. We consider the free-flyer kinematics [129] given by

$$\dot{\tilde{p}} = u_1 + w_p, \text{ and} \tag{7.17a}$$

$$\dot{\tilde{\theta}} = \Psi(\tilde{\theta})u_2 + w_\theta, \tag{7.17b}$$

where $\tilde{p} \in \mathbb{R}^3$ is the position, $\tilde{\theta} \in \mathbb{R}^3$ the attitude euler angles and $\Psi(\theta)$ the attitude Jacobian that translates body rates to Euler angles [47],

$$\Psi(\theta) = \begin{bmatrix} 1 & s_\phi t_\varphi & c_\phi t_\varphi \\ 0 & c_\phi & -s_\phi \\ 0 & s_\phi/c_\varphi & c_\phi/c_\varphi \end{bmatrix} \tag{7.18}$$

where $\phi \in]-\pi, \pi[, \varphi \in]-\frac{\pi}{2}, \frac{\pi}{2}[, \psi \in [-\pi, \pi]$ correspond to the roll, pitch and yaw angles, respectively, and $s_i = \sin i, c_i = \cos i, t_i = \tan i, i = \phi, \varphi$. Then, $\theta := [\phi, \varphi, \psi]^T$. The control input to (7.17) is a linear velocity u_1 and an angular velocity u_2, which are concatenated in $u := [u_1, u_2]^T \in \mathbb{U}$. Without loss of generality, we assume that $\mathbb{U} = \{u \in \mathbb{R}^6 : \|u_1\| \leq \bar{u}_1 \wedge \|u_2\| \leq \bar{u}_2\}$. Disturbances w_p and w_θ are upper bounded on the sets $\mathbb{W}_p := \{w_p \in \mathbb{R}^3 : \|w_p\| \leq \overline{w}_p\}$ and $\mathbb{W}_\theta := \{w_\theta \in \mathbb{R}^3 : \|w_\theta\| \leq \overline{w}_\theta\}, \forall t \geq 0$, respectively.

For the system at hand, it is convenient to separate the problem in two barriers, one for the translation and another for the attitude. Accordingly, these candidate barriers are written as

$$h_1(\tilde{p}, t) = \varepsilon_p^2 - \|\tilde{p} - p_r(t)\|_2^2, \tag{7.19}$$

$$h_2(\tilde{\theta}, t) = \varepsilon_\theta^2 - \|\tilde{\theta} - \theta_r(t)\|_2^2 \tag{7.20}$$

where $p_r : \mathbb{R}_{\geq 0} \to \mathbb{R}^3$ and $\theta_r : \mathbb{R}_{\geq 0} \to \mathbb{R}^3$ are bounded, continuously differentiable, time-varying references, with bounded derivatives. For simplicity, we drop the explicit time dependence for these variables. The barriers (7.19) and (7.20) represent a safe set around a time varying reference, where the maximum allowed error is given by $\varepsilon_p \in \mathbb{R}_{>0}$ and $\varepsilon_\theta \in \mathbb{R}_{>0}$, for the position and attitude, respectively. Proposition 2 establishes the validity of the candidate barrier h_1, while h_2 is handled in a similar way in a manner given after the proof.

Proposition 2 *Consider the kinematics in (7.17a) and the candidate barrier function h_1 in (7.19). Let $\mathcal{C}_p(t)$ be defined in (2.6) for $h := h_1$. Given a compact set*

$\mathcal{D}_p$ such that $\mathcal{C}_p(t) \subset \mathcal{D}_p, \forall t \geq 0$, let $\bar{\epsilon}_p = |\min_{\tilde{p} \in \mathcal{D}_p, t \in [0,\infty)} h_1(\tilde{p}, t)|$, and given a $\delta_p \in (0, \varepsilon_p^2)$ suppose the following holds

$$\delta_p \frac{2\bar{u}_1(\varepsilon_p^2 - \delta_p)}{c_8} - (c_6 + c_7\bar{w})(\delta_p + \bar{\epsilon}_p) > 0. \tag{7.21}$$

Then, there exists a $\Delta t^* > 0$ such that $\forall \Delta t \in (0, \Delta t^*)$, h_1 is a SD-ZCBF.

Proof: Consider the system in (7.17a) and (7.6), (7.7). Let $\alpha_1(h_1) = \lambda_1 h_1$. For (7.17a), it holds that $L_{f_c} h_1(\tilde{p}, k) = 0$ and $L_{g_c} h_1(\tilde{p}, k) = -2(\tilde{p} - p_r(k))$, $\bar{c}_2 = \underline{c}_2 = 0$, $c_4 = c_5 = \bar{u}_1$, $\bar{w} = \max_{t \in [0,\infty)} \|w_p(t)\|$. Then, let $\bar{c}_1, \bar{c}_3 \in \mathbb{R}_{>0}$ be the respective Lipschitz constants of $\lambda_1 h_1(\tilde{p}, t)$, $L_{g_c} h_1(\tilde{p}, t)$ with respect to $\tilde{p}$, $\underline{c}_1, \underline{c}_3 \in \mathbb{R}_{>0}$ be the respective Lipschitz constants of $\lambda_1 h_1(\tilde{p}, t)$, $L_{g_c} h_1(\tilde{p}, t)$ with respect to t, $\forall \tilde{p} \in \mathcal{D}_p, t \in [0, \infty)$, $c_6 = \max_{\tilde{p} \in \mathcal{D}_p, t \in [0,\infty)} \|2(\tilde{p} - p_r(t))^T \dot{p}_r(t)\|$, $c_7 = \max_{\tilde{p} \in \mathcal{D}_p, t \in [0,\infty)} \|2(\tilde{p} - p_r(t))\|$, and let $\nu(\Delta t)$ be defined as in (7.7).

Let $u_1(k\Delta t)$ be defined as

$$u_1(k\Delta t) = \begin{cases} -\frac{1}{2}(\tilde{p} - p_r(k\Delta t))\kappa_p, & \text{for } h_1 \in [-\bar{\epsilon}_p, \delta_p], \\ 0, & \text{otherwise} \end{cases}, \tag{7.22}$$

$\forall k \in \mathbb{N}_0$, where $\bar{\epsilon}_p = |\min_{\tilde{p} \in \mathcal{D}_p, t \in [0,\infty)} h_1(\tilde{p}, t)|$, $0 < \delta_p < \varepsilon_p^2$ and $\kappa_p \in \mathbb{R}_{>0}$. We note that $\bar{\epsilon}_p$ is well-defined due to $\mathcal{D}$ being compact and $p_r(t)$ bounded. Furthermore, u_1 is well defined on $\mathcal{D}$ for any $k \in \mathbb{N}_0$. We will show that with this choice of u_1 and α_1, (7.6) hold.

We will start by looking at the case when $h_1 \in [-\bar{\epsilon}_p, \delta_p]$. For this case, we substitute u_1 into the left hand-side of (7.6): $\|\tilde{p} - p_r(t)\|_2^2 \kappa_p + \lambda_1 h_1$. Considering the value of h_1 at the extremes of its interval, we can write $h_1 \leq \delta_p \implies \|\tilde{p} - p_r(t))\|_2^2 \geq \varepsilon_p^2 - \delta_p$, which together with $h_1 \geq -\bar{\epsilon}_p$ yields $\|\tilde{p} - p_r(t))\|_2^2 \kappa_p + \lambda_1 h_1 \geq (\varepsilon_p^2 - \delta_p)\kappa_p - \lambda_1 \bar{\epsilon}_p$. Furthermore, we choose κ_p to satisfy the following condition

$$\kappa_p \geq \frac{\lambda_1 \bar{\epsilon}_p + \nu(\Delta t) + c_6 + c_7\bar{w}}{\varepsilon_p^2 - \delta_p} \tag{7.23}$$

for which (7.6) holds.

Next, we investigate the case when $h_1 \in [\delta_p, \varepsilon_p^2]$. Selecting λ_1 such that

$$\lambda_1 \geq \frac{\left((\bar{c}_2 + \bar{c}_3 c_4)(c_5 + \bar{w}) + (\underline{c}_2 + \underline{c}_3 c_4)\right)\Delta t + c_6 + c_7\bar{w}}{\delta_p - (c_7(c_5 + \bar{w}) + c_6)\Delta t} \tag{7.24}$$

guarantees that (7.6) holds.

For notation brevity, let $d_1 = ((\bar{c}_2 + \bar{c}_3 c_4)(c_5 + \bar{w}) + (\underline{c}_2 + \underline{c}_3 c_4))$, $d_2 = c_7(c_5 + \bar{w}) + c_6$, $d_3 = c_6 + c_7\bar{w}$, $d_4 = \varepsilon_p^2 - \delta_p$ and $d_5 = \frac{2\bar{u}_1 d_4}{c_8}$ and note that $d_1, ..., d_5 \in \mathbb{R}_{>0}$, $\bar{c}_1 = \lambda_1 c_7$ and $\underline{c}_1 = \lambda_1 c_6$. In this manner, (7.23) and (7.24) may be written as $\kappa_p \geq \frac{\lambda_1 \bar{\epsilon}_p + \nu(\Delta t) + d_3}{d_4}$ and $\lambda_1 \geq \frac{d_1 \Delta t + d_3}{\delta_p - d_2 \Delta t}$, respectively. To ensure that $\|u_1\| \in \mathbb{U}$

and that (7.6) is satisfied, we include the actuation limit $\bar{u}_1 \geq \| - \frac{1}{2}c_8\kappa_p\|$, where $c_8 = \max_{\tilde{p}\in\mathcal{D}_p, t\in[0,\infty)} \|\tilde{p} - p_r(t)\|$. Then, we have $\bar{u}_1 \geq \kappa_p c_8 \frac{1}{2}$, which with (7.23) leads to $\frac{2\bar{u}_1}{c_8} \geq \frac{1}{d_4}(\lambda_1 \bar{\epsilon}_p + \nu(\Delta t) + d_3)$. With $\bar{c}_1 = \lambda_1 c_7$ and $\underline{c}_1 = \lambda_1 c_6$, we get $\frac{2\bar{u}_1 d_4}{c_8} \geq \lambda_1 \bar{\epsilon}_p + \lambda_1 d_2 \Delta t + d_1 \Delta t + d_3$, and with (7.24), yields $d_5 - d_3 \geq \frac{d_1\Delta t + d_3}{\delta_p - d_2\Delta t}(\bar{\epsilon}_p + d_2\Delta t) + d_1\Delta t$. At this point, let $\Delta t^* = \min\{\Delta t_1^*, \Delta t_2^*\} > 0$, where $\Delta t_1^* = \frac{\delta_p}{d_2}$ and $\Delta t_2^* = \frac{\delta_p d_5 - d_3(\delta_p + \bar{\epsilon}_p)}{\delta_p d_1 + d_2 d_5 + d_1 \bar{\epsilon}_p}$. Then, for any $\Delta t \in (0, \Delta t^*)$, the following relation holds:

$$(\delta_p - d_2\Delta t)(d_5 - d_3 - d_1\Delta t) \geq (d_1\Delta t + d_3)(\bar{\epsilon}_p + d_2\Delta t). \tag{7.25}$$

We can now expand the left and right hand-sides of (7.25) to $\delta_p d_5 - \delta_p d_3 - \delta_p d_1\Delta t - d_2 d_5\Delta t + d_2 d_3\Delta t + d_1 d_2(\Delta t)^2 \geq d_1\bar{\epsilon}_p\Delta t + d_1 d_2(\Delta t)^2 + d_3\bar{\epsilon}_p + d_2 d_3\Delta t$. Simplifying this expression and factoring out Δt yields

$$\frac{\delta_p d_5 - d_3(\delta_p + \bar{\epsilon}_p)}{\delta_p d_1 + d_2 d_5 + d_1\bar{\epsilon}_p} \geq \Delta t. \tag{7.26}$$

Thus, there exists a non-empty interval $(0, \Delta t^*)$, for which any $\Delta t \in (0, \Delta t^*)$ ensures that h_1 is a SD-ZCBF. Since there exists a u_1 such that $\|u_1\| \leq \bar{u}_1$, the candidate barrier function (7.19) is a valid SD-ZCBF, completing the proof.

$\square$

Repeating Proposition 2's proof for h_2 and considering the candidate control input

$$u_2(k) = \begin{cases} -\frac{1}{2}\Psi(\tilde{\theta})^{-1}(\tilde{\theta} - \theta_r(k))\kappa_\theta, & \text{for } h_2 \in [-\bar{\epsilon}_\theta, \delta_\theta] \\ 0, & \text{otherwise} \end{cases} \tag{7.27}$$

leads to the existence of a similar upper bound for Δt, proving that h_2 is, also, a valid SD-ZCBF.

It is important to note that there exists a trade-off between the maximum control input allowed by the system, the maximum disturbance and reference rate-of-change that the system can handle, and the maximum allowed tracking error. This relation is evident from (7.21). For small $\bar{u}_1$, ε needs to be large to satisfy $\Delta t > 0$. The same relation holds for large disturbances or fast-changing references, which influence the constants c_6, c_7 and $\bar{w}$.

Remark 6 *It is important to note that the candidate controllers (7.22) and (7.27) are only feasible solutions. With CMPC, an optimized control input is generated, respecting the constraint (7.10c) for both (7.19) and (7.20).*

7.7 Numerical Results

In this section we show simulation results for the CMPC, when applied to the free-flyer kinematics in Section 7.6. We consider the barriers defined in (7.19) and (7.20). To derive a feasible sampling time Δt that respects the conditions in Proposition 2,

Translation Dynamics - (7.17a)	Attitude Dynamics - (7.17b)
$c_4 = 0.8660$	$c_4 = 0.3464$
$c_7 = 3.4641$	$c_7 = 1.0392$
$\bar{c}_2 = \underline{c}_2 = 0$	$\bar{c}_2 = \underline{c}_2 = 0$
$\bar{c}_3 = 2$	$\bar{c}_3 = 2$
$\underline{c}_3 = 2(c_4 + 0.1)$	$\underline{c}_3 = 2(c_4 + 0.01)$
$c_5 = c_4$	$c_5 = c_4$
$c_6 = 0.1c_7$	$c_6 = 0.01c_7$
$c_8 = 0.5c_7$	$c_8 = 0.5c_7$
$\lambda_1 = 18.4374$	$\lambda_2 = 0.8584$
$\varepsilon_p = 0.9263$	$\varepsilon_\theta = 0.5095$
$\delta_p = 0.2989$	$\delta_\theta = 0.2476$
$\Delta t_p = 0.0865$	$\Delta t_\theta = 0.2337$

Table 7.1: System constants for the dynamics in (5.5).

while minimizing $\varepsilon_{p/\theta}$ and $\delta_{p/\theta}$, a nonlinear optimization routine was implemented in Matlab, taking into account the constants in Table 7.1 for the dynamics in (7.17a) and (7.17b). Accordingly, the Corridor MPC was initialized with $l(e, u) = e^T Q e + u^T R u$, $V(e) = e^T P e$, where $Q = \text{diag}([100, 100, 100, 10, 10, 10])$, $R = 50 \cdot I_6$, $P = 100 \cdot Q$, where I_6 is the identity matrix in $\mathbb{R}^{6 \times 6}$, a prediction horizon of $N = 30$ and $\Delta t = 0.07s$. Lastly, the reference trajectory was created according to Assumption 3 with $x_r(0) = 0_6$ and $u_r(t) = [0.1, 0, 0, 0, 0, 0.01]^T$, and the system initial state was set to $\tilde{x}(0) = [0.8, 0, 0, 0, 0.45, 0]$, where 0_6 is a zero vector in $\mathbb{R}^6$. The system was simulated for 15 seconds. The results for this simulation are presented in Fig. 7.2.

As it is possible to observe from Figure 7.2, the control limits are respected over the entire operation, while the system state is kept inside the safe region, delimited by the dashed lines in the control and error plots. It is also clear that the barrier values are always positive, showing that the system is kept within the prescribed safety bounds.

7.8 Discussion

In this chapter we present a novel concept for combining safety and optimality for trajectory tracking controllers, named Corridor MPC. We provide recursive feasibility and practical stability results for all states starting inside a user-defined maximum error with respect to a time-varying trajectory. As future work, possible research directions span: introducing practical stability guarantees for the sampled-data system; introducing less conservative bounds for the continuous-time safety conditions; and lastly, exploring distributed frameworks for the Corridor MPC formulation, in a multi-agent setting.

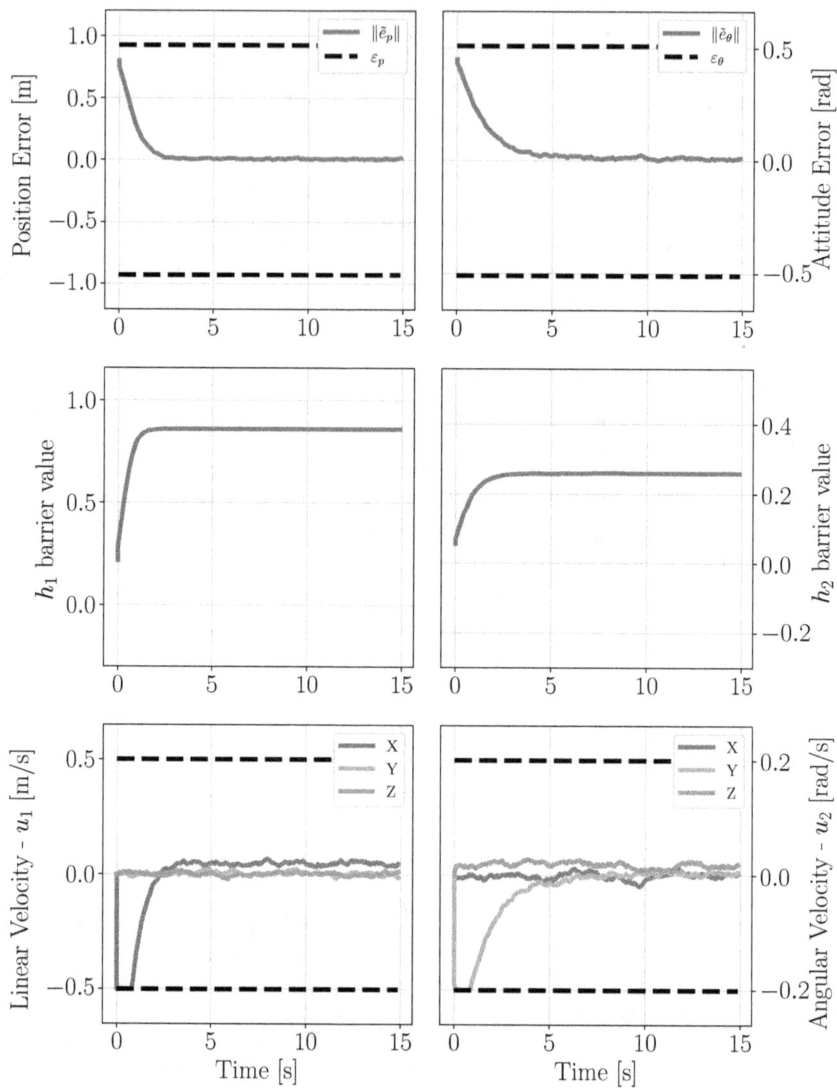

Figure 7.2: Corridor MPC for a Free-flyer Kinematics. The system is robustly kept inside a safe corridor, and stabilized on a neighborhood of the origin through optimal constrained control inputs.

Chapter 8

Conclusions and Future Work

This book focused on the problems of multi-agent formation control under limited communication and vision-based feedback. Moreover, a framework for robust and optimal control was developed that ensures safety of the agent inside a corridor that surrounds a pre-specified time-varying trajectory. These contributions were tested with realistic simulation models and in experimentally changing conditions, and each of them opens new research directions that can be further explored.

Chapter 3 dealt with formation control for a group of quadrotors under a star communication graph that aims at reducing the communication requirements to coordinate multi-agent swarms. Stability and recursive feasibility guarantees are given to the proposed model predictive controller. The goal for this framework, and its main motivation, were to coordinate SPHERE satellites in formation control demonstrations aboard the International Space Station. This work has been improved and ported to the new free-flyers aboard the space station, the NASA Astrobee, with a publication under preparation. Future work will focus on deriving formal guarantees for a tree-graph communication setting, and include obstacle avoidance in the formation control method.

Chapter 4 proposes centralized and decentralized controllers for collaborative load transportation with UAVs. The general idea proposed in Chapter 3 was used to formulate decentralized controllers for load transportation, in order to derive efficient decentralized control laws. The two approaches were compared both in simulation and in an experimental setup, providing insight into the advantages and disadvantages of each scheme. In particular, the centralized controller is computationally heavier, but is more robust to modeling errors, while the decentralized controller derives control inputs faster, but more susceptible to modeling errors. In particular, I am interested on exploring how this framework would behave when applied to space, where multiple Astrobee could work together to unload a cargo spacecraft.

In Chapter 5, vision-based control strategies are proposed to deal with setpoint stabilization of quadrotors. By using MPC, we can track a desired velocity setpoint from an image-based visual servoing module considering the UAV dynamics. Stability guarantees for the reference tracking MPC are provided. However, we cannot provide closed-loop stability guarantees for the cascaded controllers, as it is not possible to predict accurately the image-based feature motion without knowing its exact 3D position. In Chapter 6 this issue was addressed by formulating a Visual Predictive Controller, which was used as basis for the image-based formation control framework here proposed. As image-features are used in the state of the system, we can provide asymptotic stability guarantees for our controllers. Future work will expand on the formation control framework to actively shape the formation geometry to improve the depth estimation of point-features in the world, potentially leading to optimal reconstruction of the environment.

Lastly, Chapter 7 introduces a robust and safe predictive control framework based on control barrier functions, where an agent is kept inside a safe corridor around a predefined planned trajectory. Stability guarantees are provided and simulations showcase the potential of this framework for future applications. Future work spanning from the proposed Corridor MPC will deal with porting the framework to second-order systems and reducing the conservatism of the corridor guarantees, associated with the use of Lipchitz constants to upper bound the system dynamics.

As a follow-up to this book, most of the methods here proposed will be used on the new facilities proposed by the DISCOWER project for space and sub-sea applications. Particularly focusing on weightless environments will open new application directions for the methods here proposed and enhance their utility in mission-critical scenarios.